Advance Praise for

State of the World 2015: Confronting Hidden Threats to Sustainability

"The *State of the World* report has always been at the forefront of bringing new issues and new perspectives to wide public attention in the United States and abroad. Devastating findings about our world have been matched with exciting opportunities for action. For thirty years, there has been nothing quite like it, and now comes another major contribution in the series. The 2015 volume once again explores new territory, including, I am happy to see, an examination of whether GDP growth, as we now experience it in places like the United States, brings more costs than benefits. Overall, an extremely important contribution."

—**James Gustave Speth**, Author, *America the Possible: Manifesto for a New Economy*

"*State of the World* lays out the essential knowledge of what is happening to the planet and the solutions we need to sustain civilization. A cutting-edge compass for a sustainable future: no leader should be without it."

—**Thomas E. Lovejoy**, Professor of Environmental Science and Policy, George Mason University

"As the world has evolved over the past three decades, so has the annual *State of the World* report. This edition offers an even more systemic analysis than past ones, integrating finance into its snapshot of ecological and societal trends on Planet Earth. Everyone who cares about the overarching issues shaping our future should pay attention, as the warning lights on our dashboard are glowing."

—**Richard Heinberg**, Senior Fellow, Post Carbon Institute

"Each year, we can count on *State of the World* for up-to-date research revealing critical issues of sustainability for our planet and its people. Get it if you want to understand emerging threats and new possibilities. Probably the best investment you will make in 2015!

—**Mathis Wackernagel**, President, Global Footprint Network

About Island Press

Since 1984, the nonprofit organization Island Press has been stimulating, shaping, and communicating ideas that are essential for solving environmental problems worldwide. With more than 1,000 titles in print and some 30 new releases each year, we are the nation's leading publisher on environmental issues. We identify innovative thinkers and emerging trends in the environmental field. We work with world-renowned experts and authors to develop cross-disciplinary solutions to environmental challenges.

Island Press designs and executes educational campaigns in conjunction with our authors to communicate their critical messages in print, in person, and online using the latest technologies, innovative programs, and the media. Our goal is to reach targeted audiences—scientists, policymakers, environmental advocates, urban planners, the media, and concerned citizens—with information that can be used to create the framework for long-term ecological health and human well-being.

Island Press gratefully acknowledges major support of our work by The Agua Fund, The Andrew W. Mellon Foundation, The Bobolink Foundation, The Curtis and Edith Munson Foundation, Forrest C. and Frances H. Lattner Foundation, The JPB Foundation, The Kresge Foundation, The Oram Foundation, Inc., The Overbrook Foundation, The S.D. Bechtel, Jr. Foundation, The Summit Charitable Foundation, Inc., and many other generous supporters.

State of the World 2015

Confronting Hidden Threats to Sustainability

Other Worldwatch Books

State of the World 1984 through *2014*
(an annual report on progress toward a sustainable society)

Vital Signs 1992 through *2003* and *2005* through *2014*
(a report on the trends that are shaping our future)

Saving the Planet
Lester R. Brown
Christopher Flavin
Sandra Postel

How Much Is Enough?
Alan Thein Durning

Last Oasis
Sandra Postel

Full House
Lester R. Brown
Hal Kane

Power Surge
Christopher Flavin
Nicholas Lenssen

Who Will Feed China?
Lester R. Brown

Tough Choices
Lester R. Brown

Fighting for Survival
Michael Renner

The Natural Wealth of Nations
David Malin Roodman

Life Out of Bounds
Chris Bright

Beyond Malthus
Lester R. Brown
Gary Gardner
Brian Halweil

Pillar of Sand
Sandra Postel

Vanishing Borders
Hilary French

Eat Here
Brian Halweil

Inspiring Progress
Gary Gardner

State of the World 2015

Confronting Hidden Threats to Sustainability

Gary Gardner, Tom Prugh, and Michael Renner, *Project Directors*

Katie Auth
Ben Caldecott
Peter Daszak
Heather Exner-Pirot
Gary Gardner
François Gemenne
Nathan John Hagens
Tim Jackson
William B. Karesh
Elizabeth H. Loh
Catherine C. Machalaba
Tom Prugh
Robert Rapier
Michael Renner
Peter A. Victor

Lisa Mastny, *Editor*

Washington | Covelo | London

1400 16th Street, N.W.
Suite 430
Washington, D.C. 20036
www.worldwatch.org

The State of the World and Worldwatch Institute trademarks are registered in the U.S. Patent and Trademark Office.

The views expressed are those of the authors and do not necessarily represent those of the Worldwatch Institute; of its directors, officers, or staff; or of its funders.

Island Press is a trademark of Island Press/The Center for Resource Economics.

Paper:
978-1-61091-610-3
1-61091-610-7
Ebook:
978-1-61091-611-0
1-61091-611-5

The text of this book is composed in Minion, with the display set in Myriad Pro. Book design and composition by Lyle Rosbotham.

Printed on recycled, acid-free paper

Manufactured in the United States of America

Worldwatch Institute Staff

Katie Auth
Research Associate, Climate and Energy Program

Barbara Fallin
Director of Finance and Administration

Mark Konold
Research Associate and Caribbean Program Manager, Climate and Energy Program

Max Lander
Research Associate, Climate and Energy Program

Haibing Ma
China Program Manager

Lisa Mastny
Senior Editor

Donald Minor
Development Associate, Customer Relations, and Administrative Assistant

Evan Musolino
Research Associate and Renewable Energy Indicators Program Manager, Climate and Energy Program

Alexander Ochs
Director, Climate and Energy Program

Tom Prugh
Codirector, State of the World

Mary C. Redfern
Director of Institutional Relations, Development

Michael Renner
Senior Researcher

Worldwatch Institute Fellows, Advisers, and Consultants

Erik Assadourian
Senior Fellow

Robert Engelman
President Emeritus/ Senior Fellow

Christopher Flavin
President Emeritus/ Senior Fellow

Gary Gardner
Senior Fellow

Corey Perkins
Information Technology Manager

Lyle Rosbotham
Art and Design Consultant

Acknowledgments

"Teamwork divides the task and multiplies the success," the saying goes. This is profoundly true of the job of producing the annual *State of the World* report. Our work is lightened immeasurably by the dedicated efforts of countless individuals in dozens of countries, and their contributions greatly swell the book's impact and reach. All deserve our sincere thanks for their labor on behalf of the book and the Worldwatch Institute.

We are grateful to our dedicated Board of Directors for their unflagging support and leadership: Ed Groark, Robert Charles Friese, Nancy Hitz, John Robbins, L. Russell Bennett, Mike Biddle, Cathy Crain, Tom Crain, James Dehlsen, Edith Eddy, Christopher Flavin, Ping He, Jerre Hitz, Bo Normander, David Orr, and Richard Swanson, in addition to our Emeritus Directors, Øystein Dahle and Abderrahman Khene.

We add a special note of thanks to Jerre Hitz and Nancy Hitz, who are stepping down after eight years of distinguished service. Their collegial and generous approach to Board responsibilities helped make Board operations smooth and productive, and each played important roles in setting the Institute's direction at critical junctures. We extend to them our deepest thanks for their dedication and service, along with sincere wishes for great success in the years ahead.

Thank you as well to the many institutional funders whose support made Worldwatch's work possible over the past year. We are grateful to (in alphabetical order): the Ray C. Anderson Foundation; Asian Development Bank; Charles and Mary Bowers Living Trust; Carbon War Room Corporation; Caribbean Community (CARICOM); Climate and Development Knowledge Network; Cultural Vision Fund of the Orange County Community Foundation; Del Mar Global Trust; Doughty Hanson Charitable Foundation; Eaton Kenyon Fund of the Sacramento Region Community Foundation; Embassy of the Federal Republic of Germany to the United States; The Friese Family Fund; Garfield Foundation; Brian and Bina Garfield, Trustees; German Federal Ministry for the Environment, Nature Conservation and Nuclear Safety (BMU) and the International Climate Initiative; William and Flora Hewlett Foundation with Population Reference Bureau; Hitz Foundation; Inter-American Development Bank; Steven Leuthold Family Foundation; National

Renewable Energy Laboratory, U.S. Department of Energy; Renewable Energy Policy Network for the 21st Century (REN21); MAP Royalty, Inc., Natural Gas and Wind Energy Partnerships; Mom's Organic Market; Network for Good; Quixote Foundation, Inc.; Randles Family Living Trust; V. Kann Rasmussen Foundation; Estate of Aldean G. Rhyner; Serendipity Foundation; Shenandoah Foundation; Flora L. Thornton Foundation; Turner Foundation, Inc.; United Nations Foundation; United Way of Central New Mexico; Johanette Wallerstein Institute; Wallace Global Fund; Weeden Foundation Davies Fund; and World Bank International Finance Corporation with CPCS Transcom Ltd.

Support from thousands of Friends of Worldwatch strengthens the Institute's budgetary position and provides stability to our financial planning. This core group of readers and donors is deeply loyal to the Institute and dedicated to creating a sustainable civilization, and provides an indispensable and stable financial base, year after year, for our work.

For our thirty-second edition, the Institute welcomes submissions from a wide range of authors—all of whom contribute atop the many pressures of their own work. We are grateful to Katie Auth, Ben Caldecott, Peter Daszak, Heather Exner-Pirot, François Gemenne, Nate Hagens, Tim Jackson, William B. Karesh, Elizabeth H. Loh, Catherine C. Machalaba, Robert Rapier, and Peter Victor for their generous contributions to the book. Seldom has the affirmation "we couldn't have done it without you" been so literally true.

State of the World is ably edited by Lisa Mastny, whose skill at converting the language of diverse authors into clear and consistent prose makes the book accessible to a broad audience. Lisa is a nimble and highly organized manager who coordinates the work of dozens of authors and others to meet a firm deadline. She is also consistently cheerful and exceedingly diplomatic, which makes the production process surprisingly painless. Lyle Rosbotham uses his exceptional design talents to turn printed words into engaging text and graphics in a beautiful layout. We are grateful to Lyle for his continued involvement in making the book as engaging as it is. We also thank Kate Mertes for her work in preparing the index.

Producing and printing *State of the World* is just the beginning of the project effort. Communications Director Gaelle Gourmelon works to ensure that the book's messages reach far beyond our Washington offices and offers key input to project design decisions. Director of Finance and Administration Barbara Fallin ensures that the trains run on time at Worldwatch, through her efficient management of daily operations. And Mary Redfern manages our relationships with foundations and other institutional funders, helping to match Institute needs with funder opportunities.

We continue to benefit from a fruitful partnership with our publisher, Island Press, which is globally recognized as a first-rate sustainability publishing house. We appreciate the professional and collegial efforts of Emily Turner Davis, Maureen Gately, Jaime Jennings, Julie Marshall, David Miller, Sharis Simonian, and the rest of the IP team.

Our network of publishing partners extends our global reach through their work in translation, outreach, and distribution of the book. We give special thanks to Universidade Livre da Mata Atlântica/Worldwatch Brasil; Paper Tiger Publishing House (Bulgaria), China Social Science Press; Worldwatch Institute Europe (Denmark); Gaudeamus Helsinki University Press (Finland); Organization Earth (Greece); Earth Day Foundation (Hungary); Centre for Environment Education (India); WWF-Italia and Edizioni Ambiente; Worldwatch Japan; Korea Green Foundation Doyosae (South Korea); FUHEM Ecosocial and Icaria Editorial (Spain); Taiwan Watch Institute; and Turkiye Erozyonla Mucadele, Agaclandima ve Dogal Varliklari Koruma Vakfi (TEMA) and Kultur Yayinlari IsTurk Limited Sirketi (Turkey).

We are also grateful to those individuals who work hard to advance the book's prospects, typically on a volunteer basis. Matt Leighninger of the Deliberative Democracy Consortium was especially helpful in responding to information requests; Stacey Rosen at the U.S. Department of Agriculture's Economic Research Service offered keen advice on Chapter 5; Gianfranco Bologna continues to promote *State of the World* in Italian, and for 20 years has been a gracious host in Italy; Eduardo Athayde is an irrepressible font of ideas and promoter of Worldwatch in Brazil; and Japan's Soki Oda casts a keen critical eye on Worldwatch research as perhaps the most meticulous reader of our work.

Finally, we extend a special note of thanks to Bob Engelman, who stepped down as President of the Institute in 2014, and to Ed Groark, who stepped in as Acting Interim President. Bob provided a steady hand over several years of transition, and we grew to admire his judgment, encouragement, and leadership. Ed's unshakable belief in the Worldwatch mission, his creative thinking, and his untiring efforts are raising the Institute to a new level of excellence. In a world of flux in the publishing and nonprofit realms, the Institute has been fortunate to have Bob and Ed at the helm.

Gary Gardner, Tom Prugh, and Michael Renner
Project Directors
Worldwatch Institute
1400 16th St. N.W., Suite 430
Washington, D.C. 20036
worldwatch@worldwatch.org
www.worldwatch.org

Contents

BOXES

TABLES

FIGURES

Units of measure throughout this book are metric unless common usage dictates otherwise.

Introduction

CHAPTER 1

The Seeds of Modern Threats

Michael Renner

On September 21, 2014, an estimated 400,000 people marched in New York City to demand that government leaders assembling in that city for a "climate summit" finally move from rhetoric to action. It was the largest of more than 2,600 protest events worldwide. The marches were the culmination of decades of growing climate activism that got its start soon after Dr. James Hansen put climate change on the political map. On a fittingly sweltering day in June 1988, Hansen—then the director of NASA's Goddard Institute for Space Studies—testified before the U.S. Senate's Energy and Natural Resources Committee that global warming was not a natural phenomenon, but rather was caused by human activities that triggered a buildup of greenhouse gases in the atmosphere.[1]

Hansen was far from the first scientist to theorize about human-induced climate change. Such studies go back as far as the late nineteenth century, but by the 1960s and 1970s, scientists started to view the warming potential of gases like carbon dioxide as increasingly convincing. In February 1979, the World Meteorological Organization (WMO) concluded in its "Declaration of the World Climate Conference" that "it appears plausible that an increased amount of carbon dioxide in the atmosphere can contribute to a gradual warming of the lower atmosphere. . . . It is possible that some effects on a regional and global scale may be detectable before the end of this century and become significant before the middle of the next century." By the 1980s, the pace of climate studies quickened, and the Intergovernmental Panel on Climate Change (IPCC) was set up in 1988 by the WMO and the United Nations Environment Programme (UNEP).[2]

It was Hansen, however, who conveyed an unmistakable sense of urgency, telling the assembled senators in 1988: "It's time to stop waffling so much and say that the evidence is pretty strong that the greenhouse effect is here." Yet his testimony marked merely the beginning of a protracted struggle to get governments, corporations, and society at large to understand that

Michael Renner is a senior researcher at the Worldwatch Institute and codirector of *State of the World 2015*.

NASA

Above: James Hansen testifying in 1988.

Right: Hansen getting arrested at a civic protest in 2011.

Ben Powless

humanity's own actions have brought about a challenge unlike any other—and then to act on that understanding.[3]

During the past quarter century, much has indeed changed. From Hansen's early findings, climate modeling became ever more sophisticated, observational work multiplied, and scientific consensus solidified. The world's governments came together in 1992 and set up the United Nations Framework Convention on Climate Change, the starting shot for a process of annual "conferences of the parties" (COPs) charged with negotiating a global climate treaty. Climate change, once the preserve of very few specialists, has become a household word. The number of studies and reports on climate impacts and possible solutions has mushroomed. By late 2013, the IPCC concluded that it "is extremely likely that human influence has been the dominant cause of the observed warming since the mid-twentieth century."[4]

However, lofty rhetoric has far outpaced action. Climate negotiations have failed to deliver anything close to the breakthrough agreement that the world desperately needs. Hansen's own sense of increasing urgency moved him from scientific inquiry toward activism in recent years. He was even arrested a few times at high-profile civic protests.

Strangely, we now find ourselves in an era of "sustaina*babble*"—marked by wildly proliferating claims of sustainability. Even as adjectives like "low-carbon," "climate-neutral," "environmentally friendly," and "green" abound, there is a remarkable absence of meaningful tests for whether particular governmental and corporate actions actually merit such descriptions.[5]

Meanwhile, powerful fossil fuel interests have mobilized with great effectiveness to thwart action amid all this hot air, sowing doubt and confusion about climate science, and opposing or delaying effective policy making. It brings to mind a quote from author Upton Sinclair, who once exclaimed that, "It is difficult to get a man to understand something, when his salary depends upon his not understanding it!"[6]

Endless economic growth driven by unbridled consumption is so central to modern economies and is so ingrained in the thinking of corporate and political leaders that environmental action is still often seen as in conflict with the economy, and is relegated to inferior status. We have an economic system that is the equivalent of a great white shark: it needs to keep water moving through its gills to receive oxygen, and dies if it stops moving. The challenge, therefore, is broader than merely a set of technological changes. As activist Naomi Klein has argued, saving the climate requires revisiting the central mechanisms of the world's pre-eminent economic system: capitalism.[7]

Shying away from such radical change, governments and international agencies are lining up behind "green growth"—a concept that reaffirms the centrality of economic growth and avoids any critique of the underlying dynamics that have brought human civilization to the edge of the abyss. According to the Organisation for Economic Co-operation and Development (OECD), "green growth means fostering economic growth and development while ensuring that natural assets continue to provide the resources and environmental services on which our well-being relies."[8]

Humanity's climate predicament is only the latest—if by far the most challenging—manifestation of its collision course with planetary limits. Ecological stress is evident in many ways, from species loss, air and water pollution, and deforestation to coral reef die-offs, fisheries depletion, and wetland losses. The planet's capacity to absorb waste and pollutants is increasingly taxed.

The *Millennium Ecosystem Assessment* found that even a decade ago, more than 60 percent of the world's major ecosystem goods and services were degraded or used unsustainably. Some 52 percent of commercial fish stocks are now fully exploited, about 20 percent are overexploited, and 8 percent are depleted. The number of oxygen-depleted dead zones in the world's oceans that cannot support marine life has doubled each decade since the 1960s; in 2008, there were more than 400 such zones, affecting an area equivalent in size to the United Kingdom. The decline of bees and other pollinators is jeopardizing agricultural crops and ecosystems. Urban air pollution causes millions of premature deaths each year. The World Health Organization recently revised its estimates of global deaths from air pollution to about 7 million people in 2012—more than double previous estimates and making air pollution the world's single worst environmental health risk.[9]

A Double-edged Sword

How did we get to this moment in time? The onset of agriculture was the first major marker of humanity's rising claim on the planet's resources, followed by the Industrial Revolution starting in the late eighteenth century. According to environmental historian J. R. McNeill, shifting agriculture improved caloric intake and thus increased energy availability perhaps 10-fold over what was available to hunter-gatherer societies. Settled agriculture provided another 10-fold increase, and domesticated animals (oxen, horses, etc.) offered concentrated muscle power for transport and plowing of fields. These were the beginnings of an—albeit still modest—energy surplus.[10]

It was the Industrial Revolution that increased that surplus beyond anything seen before, and that allowed humans to dominate Earth's biophysical systems. The invention of the steam engine permitted industrializing societies to tap coal as the primary energy source, replacing and augmenting the muscle power of humans and their domesticated animals. By 1900, steam engines had become 30 times as powerful as the first machines of around 1800. Then, by the late nineteenth century, internal combustion engines made their appearance, more efficient and powerful than steam engines, allowing for the generation of electricity and offering a means of mass transport.[11]

The period since the advent of the Industrial Revolution has seen astonishing scientific and technical advances. Whereas just 10 scientific journals were published in the mid-1700s, today they number in the tens of thousands, with estimates ranging from 25,000 to 40,000. Perhaps some 50 million scientific articles have been published since the beginning of the Industrial Revolution, with an estimated 1.4 million to 1.8 million articles published annually. Although hard to measure, one study estimated that scientific publications may be growing at an annual rate of 8–9 percent, up from just 2–3 percent during the period from the mid-eighteenth century to 1945, and less than 1 percent prior to the middle of the eighteenth century.[12]

The second half of the twentieth century, in particular, ushered in an unprecedented degree of progress in many fields, with tremendous gains in health, food availability, material well-being, and life spans. Yet these advances came at great cost to the planet's ecosystems and resources. Technical advances were often pursued single-mindedly, with little sense of restraint or long-term wisdom that might consider the repercussions for the natural world. Science, in other words, is a double-edged sword: it underpins the breathtaking progress that modern societies now take for granted, but it also enables the process that turns every last resource of the planet into a commodity.[13]

To a large extent, this is the result of large evolutionary forces—the genetic, developmental, and cultural factors that influence and determine human behavior. Humanity's ability to marshal the earth's resources, along with the economic and political competition that drives governments, corporations, and individuals, has meant that there have been few—if any—constraining factors on human actions. This lack of constraint may be the biggest threat to human survival. As J. R. McNeill observed, "The same characteristics that underwrote our long-term biological success—adaptability, cleverness—have lately permitted us to erect a highly specialized fossil fuel-based civilization so ecologically disruptive that it guarantees surprises and shocks."[14]

The industrial era's innumerable discoveries and inventions were underwritten by cheap and plentiful fossil energy. Humans used perhaps 10 times as much energy during the twentieth century as they did in the 1,000 years before. Coal, oil, and natural gas not only pack far more energy than traditional sources like wood, but their versatility allows them to be used for many different purposes, such as heating and cooling, industrial processes, electricity, and diverse forms of transport.[15]

World coal extraction shot up from about 10 million tons in 1800 to 762 million tons by 1900. It reached 4,700 million tons in 2000, and then climbed to almost 7,900 million tons in 2013—a more than 10-fold increase since 1900. World oil production started only in the late nineteenth century, but grew rapidly from 20 million tons in 1900 to 3,260 million tons in 2000, and to 4,130 million tons in 2013—a 207-fold expansion since 1900.[16]

Pre-industrial societies relied on a limited range and quantity of materials, with wood, ceramics, cotton, wool, and leather playing major roles. By contrast, industrialized societies use tens of thousands of versatile materials drawn from the entirety of naturally occurring elements. Materials like plastics or aluminum are ubiquitous nowadays (generating convenience as much as pollution), but they had their beginnings only in the late nineteenth century.[17]

Metals have long been used by humans, but their application on a mass scale is a relatively recent phenomenon. World metals production rose from 30 million tons in 1900 to 198 million tons in 1950. After reaching 740 million tons in 1974, output leveled off for the next 20 years. But then came another phase of rapid growth, driven principally by economic expansion in China, and production reached 1.7 billion tons in 2013. (See Figure 1–1.) The bulk of this figure is accounted for by steel production, which expanded 55.8-fold since 1900 and 8-fold since 1950. Aluminum production grew 32-fold since 1950, copper and zinc 6- to 7-fold, and lead and gold about 3-fold.[18]

Figure 1–1. World Metals Production, 1950–2013

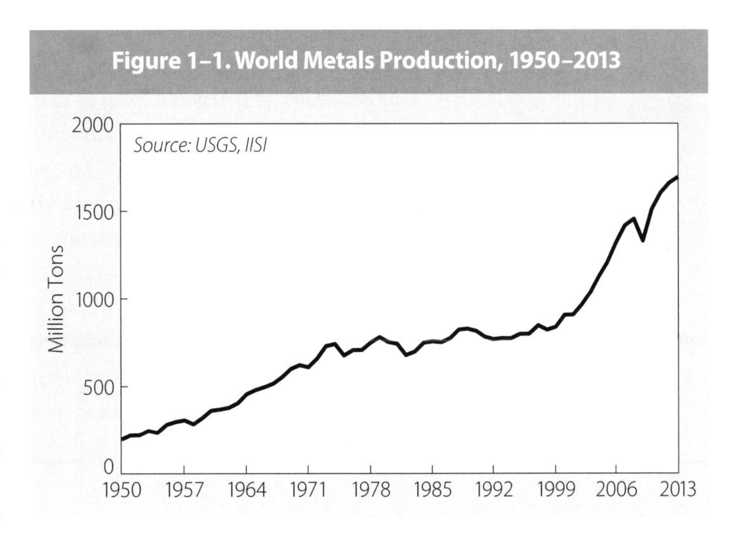

Chemical compounds have become ubiquitous to the point that a 2013 UNEP report noted, "There is hardly any industry where chemical substances are not used, and there is no single economic sector where chemicals do not play an important role." Roughly 10 million chemical compounds have been synthesized since 1900, with some 150,000 or so put to commercial use—although nobody knows the exact number. The global chemical industry's output climbed from $171 billion in 1970 to over $4.1 trillion in 2010 (expressed in nominal dollars). World chemical sales more than doubled during just the last decade, again due mostly to China, where output nearly tripled.[19]

New chemicals keep getting introduced into commerce each year—an average of 700 in the United States alone. The rising number of compounds, their increasing complexity, and an ever more intricate supply chain is giving rise to concerns that poor management of chemicals could pose substantial dangers to the health of communities and ecosystems. The industry is a perfect example of the mix of benefits and hidden threats that is so characteristic of the modern age.[20]

Increased use of synthetic fertilizers has been a key aspect of today's industrialized agriculture (along with high energy and water use and inputs like pesticides). In 1940, the world used about 4 million tons of fertilizer. By 2000, the figure reached 137 million tons, and by 2013, about 179 million tons. As J. R. McNeill reminds us, without fertilizers, "the world's population would need about 30 percent more good cropland." Massive use of synthetic fertilizers led to widespread water pollution. It also helped consolidate food production to a limited number of crops that responded well to applications of fertilizer, leading to widespread monocultures. And fertilizer production is highly energy intensive, part of the industrialization of agriculture.[21]

One of the areas in which the consequences of industrialization show up most dramatically is air quality. For most of human history, air pollution was of a local and limited nature, but during the twentieth century, it grew exponentially as heating, power generation, metal smelting, motorized transportation, waste incineration, and other human activities mushroomed.

Automobiles provide extraordinary individual mobility, but they have been a major contributor to urban air pollution. From fewer than 10,000 in 1900, 8 million cars rolled off the world's assembly lines in 1950, a number that skyrocketed to 85 million in 2013. From perhaps 25,000 cars on the world's roads in 1900 and less than 1 million in 1910, the global automobile fleet was close to 100 million in 1960 and crossed the 1 billion threshold in 2013.[22]

Pollution Control and New Growth Impulses

Massive air pollution was one of the signature issues for a budding modern environmental movement in the early 1970s, which eventually prodded governments in industrialized countries to adopt pollution control measures and to compel industry to develop more-efficient production technologies. In the United States, sulfur dioxide emissions were cut by 83 percent between 1970 and 2013, carbon monoxide emissions declined by 64 percent, nitrogen oxides by 51 percent, and volatile organic compounds by 49 percent. Better controls and more-efficient technologies also helped reduce emissions of metals like copper and lead, although they remained far above the levels of a century earlier. (See Table 1–1.)[23]

Table 1–1. World Metal Emissions to the Atmosphere, 1901–1990

Period	Cadmium	Copper	Lead	Nickel	Zinc
		annual average, in thousands of tons			
1901–1910	0.9	5.3	47	0.8	39
1951–1960	3.4	23	270	14	150
1971–1980	7.4	59	430	42	330
1981–1990	5.9	47	340	33	260

Source: See endnote 23.

During the final quarter of the twentieth century, pollution control, greater efficiency, and a degree of material saturation in the Western economies slowed further growth of production and consumption. But since the 1990s, globalization and the rise of China and a number of other "emerging economies" provided a whole new impulse for industrial development and resource use. A rising middle class in these nations started to imitate Western lifestyles, and industrial production relocated increasingly to these countries. China alone now accounts for just under half of the world's steel production, up from only 5 percent in 1980 (when worldwide production was less than half of what it is now).[24]

The 1992 Earth Summit in Rio de Janeiro was a milestone in global environmental consciousness. Yet in the two decades since then, the pressures on the planet's natural resources and ecological systems have only increased, and the second Rio conference—"Rio+20" in 2012—was far less of an environmental milestone. (See Table 1–2.) The production of energy-intensive materials—cement, plastics, and steel—has more than doubled since 1992, far outstripping overall economic growth. Global resource extraction—of fossil fuels, metals, minerals, and biomass—grew 50 percent in the 25 years between 1980 and 2005, to about 58 billion tons of raw materials (and another 40 billion tons of material removed simply to gain access to coveted resources).[25]

Recognizing and Acting on Unexpected Threats

Being science-based, modern societies eventually come to learn about the unexpected and sometimes unintended consequences of turning ever-greater portions of the planet's natural base into commodities. We have gradually come to comprehend that we are depleting resources at unsustainable

Table 1–2. Social, Economic, and Environmental Trends Between the First and Second Rio Earth Summits

	Trends	Percent Change, 1992–2012
Population and Economy	Urban population	26
	World gross domestic product (GDP)	75
	World GDP per capita	39
	World trade	311
Food and Agriculture	Food production index	45
	Irrigated area	21
	Land under organic farming	240
	Proportion of fish stocks fully exploited	13
Industry	Cement production	170
	Steel production	100
	Electricity production	66
	Plastics production	130
Transportation	Passenger car production	88
	Passenger car fleet	73
	Air transport, passengers	100
	Air transport, freight	230
Atmosphere	Carbon dioxide emissions	36
	Use of ozone-depleting substances	-93

Source: See endnote 25.

rates, spreading dangerous pollutants, undermining ecosystems, and threatening to unhinge the planet's climate balance.

But a reckoning is complicated by the fact that the complete environmental impacts of human actions are not always readily discernible. Environmental change takes place not in linear, predictable ways that can be studied in isolation from other factors, but rather entails unexpected discontinuities, synergisms, feedback loops, and cascading effects. (See Table 1–3.) And these phenomena can also reinforce each other—i.e., feedback loops can generate discontinuities, discontinuities can produce synergisms, and synergisms can trigger cascading effects. Thus, the full costs of modern conveniences often remain hidden, sometimes making themselves felt only years or even decades down the road.[26]

Table 1–3. Types of Unexpected Environmental Change

Type of Change	Definition
Discontinuity	An abrupt shift in a trend or change from a previously stable state. *Example:* Overfishing leading to a sudden crash in fish populations, rather than to a gradual decline.
Synergism	A change in which two or more phenomena combine to produce an effect that is greater than the sum of separate individual impacts. *Example:* Flood impacts magnified by a combination of deforestation and population growth in areas vulnerable to flooding.
Feedback loop	A cycle of change that amplifies itself. *Example:* Dwindling Arctic ice due to climate change causes the ocean to warm more rapidly, which in turn accelerates the loss of ice.
Cascading effects	Effects that occur when a change in one component of a system produces change in another component, which in turn changes yet another component, and so on. *Example:* A decline in herring populations depresses sea lion and seal populations, which leads killer whales to prey more on otters instead. The resulting collapse of otter populations triggers an explosion in sea urchins (the favorite prey of otters), but demolishes the kelp forests on which they feed and jeopardizes other marine species.

Source: See endnote 26.

The massive snowfall in the northeastern United States in November 2014 is just one recent illustration of such complex interactions. Rapid disappearance of Arctic sea ice north of Scandinavia due to warming temperatures leads the ocean underneath to absorb more of the sun's energy during the summertime. In the fall, the absorbed heat is released back into the atmosphere and disrupts the circumpolar winds whose patterns determine much of the weather across the earth's northern hemisphere. Scientists have found that the bubble of warm air creates a northward bulge in the jet stream. That in turn creates a surface high-pressure area circulating clockwise and pulling cold air from the Arctic over northern Eurasia—which creates a southward dip in the jet stream. The northward bulge of the jet stream over Scandinavia and the southward dip over Asia combine to create a pattern that sends energy up into the stratosphere and disrupts the polar vortex. As a consequence, frigid Arctic air gets pushed south, creating perfect conditions for massive snowfall. Scientists think that the disruptions to the jet stream and the polar vortex will become more frequent in the future, as greenhouse gas emissions continue to increase.[27]

Matters only get more complicated once scientific discovery of such environmental repercussions has taken place. Scientific findings need to be translated into a roadmap for society—into do's and don'ts. It is one thing, for instance, to stipulate that society should heed the precautionary principle (which holds that if a particular action is suspected of causing harm, the burden of proof that it is *not* harmful falls on those taking the action). But it is quite another to make society actually live by it. Resistance to needed change is not surprising in instances where local (i.e., someone else's) air or water quality is at stake, or where a given species might face extinction. After all, humans have amply demonstrated a willingness to sacrifice the well-being of certain, other, groups of people (or of animals, etc.) in exchange for short-term gain.

With civilization itself hanging in the balance, however, change in the face of climate chaos should be a no-brainer. Yet the politics of climate change to date indicates just how limited society's willingness to act on scientific advice can be. The political process through which this has to be accomplished is inevitably difficult, given that almost no aspect of human society remains untouched by efforts to stabilize the climate. But it has become more difficult in recent years by the increasing influence of money on electoral and legislative processes. In the battle to do what is needed to ensure humanity's long-term survival, a combination of denial, short-term thinking, profit interests, and human hubris is proving hard—perhaps even impossible—to overcome.

Getting society to acknowledge and address environmental and health impacts has never been an easy task. Consider these examples:

- *Leaded gasoline.* Lead was deliberately added to gasoline from the 1920s onward, after a chemical engineer discovered that it improved engine performance. Even though there were early concerns, the major proponents of this practice in the United States, General Motors and DuPont, succeeded in preventing regulations for decades. By the 1960s and 1970s, medical research showed that leaded gasoline had contributed to elevated levels of lead in people's blood. The Soviet Union first banned the practice in 1967. The United States phased out leaded gasoline in the late 1970s, Japan and Western Europe in the late 1980s, and many other countries in the 1990s. In some countries, like the United States, the fact that catalytic converters—devices introduced to reduce emissions of hydrocarbons and carbon monoxide—function properly only with lead-free fuel helped greatly in bringing about a policy change. By

2011, lead had been removed from gasoline in at least 175 countries, permitting a 90 percent drop in blood lead levels worldwide and saving an estimated 1.2 million lives each year.[28]

- *Photochemical smog.* This brown haze afflicting many cities can inflame people's breathing passages and decrease lung capacity, as well as affect the health of crops and forests. Beyond the impacts of individual air pollutants, smog is a synergistic effect that results from a cocktail of substances, including ground-level ozone, sulfur dioxide, nitrogen oxides, and carbon monoxide. It was first identified in the early twentieth century, when coal burning in cities was ubiquitous (as it still is in Chinese cities today), whereas the "modern" form of smog derives from vehicular and industrial emissions and became a problem from the 1950s on. Air pollution control measures and cleaner motor vehicle fuels have somewhat alleviated the situation, although smog continues to be a health problem in many cities around the world.[29]

Steve Jurvetson

An antenna tower in the smog of Shanghai, 2007.

- *Ozone-destroying CFCs.* A class of chemicals called chlorofluorocarbons (CFCs) was initially highly prized, given the versatile use of these compounds as refrigerants, propellants, flame retardants, and solvents. But from the mid-1970s on, scientific evidence began to mount that CFCs harm the earth's ozone layer, which protects people, animals, and plants from dangerous ultraviolet radiation. By the mid-1980s, with a dramatic seasonal depletion of the ozone layer over Antarctica, governments finally acted. The Montreal Protocol on Substances that Deplete the Ozone Layer, adopted in 1987, led to a dramatic decline in CFC use—a drop of 96 percent by 2005. A September 2014 UNEP report found that the ozone layer is slowly healing and likely will recover by mid-century. However, there is a hidden threat surrounding the hydrofluorocarbons (HFCs) that came into use as substitutes for CFCs; given that HFCs are potent greenhouse gases, safer alternatives need to be developed.[30]
- *Superbugs.* The livestock sector is characterized increasingly by industrial methods that confine animals in cramped conditions, and that

administer heavy doses of antibiotics to speed animal growth and reduce the likelihood of disease outbreak. In the United States, almost four times the amount of antibiotics is used in livestock operations as to treat ill people. However, such indiscriminate practices pose a threat to antibiotics' effectiveness for human uses—one that has been recognized widely, but not acted on. A similar problem lies in the overuse of herbicides and pesticides, as well as the development of genetically modified seeds that emit their own pesticide. As insects develop resistance to such products, farmers confront the danger of catastrophic harvest failure.[31]

Climate change is multiplying these kinds of problems. Humanity is ever so slowly coming to grips with the growing reality of a destabilized climate. Even as scientists and others shed light on the likely repercussions such as sea-level rise, droughts, floods, and superstorms, some challenges remain undetected or at least underappreciated. These challenges—several of which are discussed in the chapters that follow—concern not only environmental dynamics themselves, but also how they translate into the social, economic, and political spheres.

Energy, credit, and the end of growth. The prosperous economies and the culture of growth that industrialized nations take for normal, and that most other nations aspire to, rest on cheap (mainly fossil) energy. But, as Chapter 2 explains, we already have tapped the easy energy stores, so the push for continued growth is taking increasing amounts of energy and investment money, leaving less for every other activity. Moreover, the thousands of energy "slaves" we each have working for us are walking a tightrope: energy must be costly enough to be profitable for producers, yet cheap enough to be affordable to consumers. The higher that prices must rise to sustain production, the more likely is a situation of reduced demand, economic malaise, and rising debt.

Curbing growth. Economic growth drives most environmental problems, and it has produced a world in which human activities have grown too large for the planet to accommodate them sustainably. Forests are scalped, rivers run dry, species are going extinct, and humans are changing the climate, all driven by the pursuit of growth. Yet few recognize that growth itself needs to be abandoned as a national goal. Growth is widely regarded as inevitable and indispensable, but as a matter of national policy, it is barely 50 years old. Fortunately, as the authors of Chapter 3 argue, a move toward an economy that is not driven by growth of material throughput—yet that still offers adequate employment, and reduces inequality and environmental impact—is achievable.

Stranded assets. Continued investments in a fossil fuel-centered energy system—and especially in such forms of "extreme energy" as tar sands, Arctic oil deposits, shale oil and gas, and mountaintop-removal coal—will lock societies onto a dead-end path. Scientists are warning that the bulk of the world's proven fossil fuel resources can never be touched if the world wants to avoid runaway climate change. Further investing in them—and thus enlarging the carbon "bubble"—exposes not only energy companies and fossil fuel exporters to incalculable risk (a problem analyzed in Chapter 4), but also pension funds, municipal authorities, and others who invest in such companies for long-term financial returns. Absent alternative policies, the world may confront an unpalatable choice between climate chaos and economic doom.

Declining harvests. Loss or degradation of key agricultural resources—especially land, water, and a stable climate—is leading to a global agricultural system in which more countries depend on international markets for basic food supplies. Chapter 5 argues that a food import strategy reduces pressure on agricultural resources, especially water, in many countries, but also renders importing countries vulnerable to supply disruptions caused by poor harvests, political manipulation, or other factors beyond their control.

Ron Nichols, USDA NRCS

A drought-stunted soybean plant withers in the Arkansas summer sun.

Decline of the oceans. Most humans spend little time in or on the oceans, but our lives are profoundly shaped by their condition. That condition is increasingly dire. Overfishing is compromising the oceans' ability to supply the protein on which roughly 3 billion people depend. Ocean waters also function as a major sink for human-caused carbon emissions and the heat they trap in the atmosphere, but the rate of absorption of both heat and emissions may be slowing. And carbon absorption is changing the acidity of ocean waters, which in turn imperils vital marine organisms and even the marine food web itself. Chapter 6 considers these dangers.

Arctic changes. The Arctic is a showcase for the effects of climate change, especially with the alarming decline in the extent of summer sea ice and its positive feedback effects on warming. The region is an area of contention as well, as the expansion of open water entices Arctic nations with the prospect of easier access to oil and other resources. But, as Chapter 7 explores, nearly unnoticed is the struggle of Arctic peoples to ensure that the fate of the region they call home is largely in their hands, not in those of southerners seeking to impose their own political agendas.

Emerging diseases from animals. Human activities disrupt ecological systems worldwide, increasing the likelihood that infectious disease will spread from animals to humans, as has already occurred with the Ebola virus and HIV/AIDS. Scientists estimate that more than 60 percent of the 400 new infectious diseases in humans that emerged in the past 70 years were of animal origin. And this threat is increasing as land-use changes bring animals and humans together, as livestock raising becomes intensified, and as the use of antibiotics in animals increases. Chapter 8 contends that, despite rising attention to high-profile pandemics like Ebola, neither governments nor publics appreciate that such outbreaks are emblematic of a systemic, global problem.

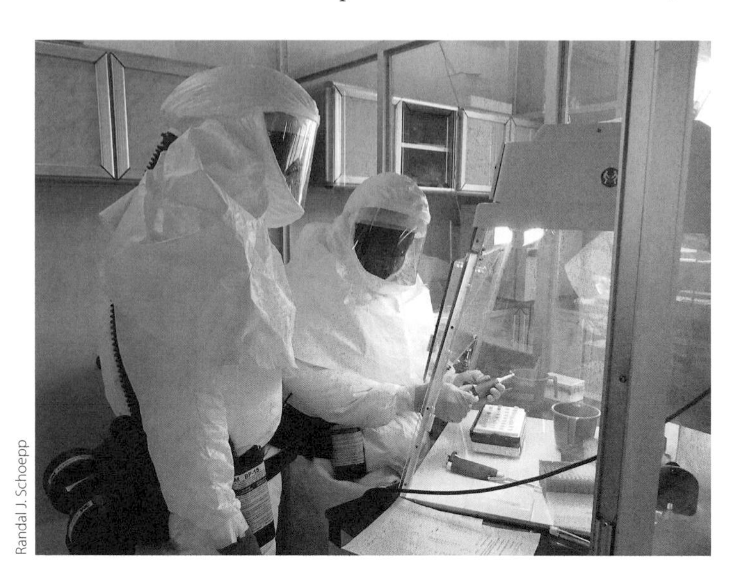

Randal J. Schoepp

U.S. Army technicians set up an assay for Ebola in a containment laboratory.

Climate migrants. Finally, population displacements due to climate change and other adverse environmental developments could undermine the social fabric of affected societies as well as trigger growing competition over resources, jobs, and social services in receiving areas. The speed, direction, and extent of such population movements remain largely the stuff of conjecture today, but they could have deeply destabilizing economic and political consequences in the future. Chapter 9 argues that timely adaptation measures—including support for migrants as well as for those who lack the resources to move—can help individuals and societies at large cope with the repercussions of a changing climate.

Conclusion

Human ingenuity has fashioned technically advanced societies and maximized the production of goods and services. Our economic systems are programmed to squeeze ever more resources from a planet increasingly in distress—whether it be more oil and gas from underground deposits, more milk from a cow, or more economic surplus from the human workforce. Although the discussion of political systems often revolves around lofty ideas like freedom, democracy, and different forms of representation, at base, they are engineered to support the process of maximizing material throughput.

But this success has come at the expense of weakened biological diversity and compromised natural systems. And it is the result of a relatively narrow set of factors and circumstances, ranging from natural conditions to human institutions. Yet these very circumstances could one day be swept away by the severe shocks that a destabilized climate entails, putting in question the ability of societies not just to thrive, but to adapt and possibly even survive. This is especially the case if societies fail to recognize hidden threats in a timely manner.

The very pillars of contemporary success—among them, high degrees of specialization, complexity, and manifold interconnections—could very well turn out to be humanity's Achilles heel. Specialization works well only within certain tightly controlled parameters, but it could be useless under changed circumstances. Complexity and interconnections multiply the strengths and advantages of a viable system, but they also make it susceptible to a rapid cascade of destabilizing impacts. Such a highly productive system is actually low on resilience because it focuses on constantly reducing any slack or redundancy—the exact features that allow for resilience to materialize. Author Thomas Homer-Dixon quotes Buzz Holling, a leading Canadian ecologist, who has warned that the longer a system is locked onto a trajectory of unsustainable growth, "the greater its vulnerability and the bigger and more dramatic its collapse will be."[32]

Seen through this broader lens, it is clear that the challenge for humanity today is no longer anything like what it faced in the 1960s and 1970s, when developing pollution abatement technologies and lessening the degree to which resources were wasted provided a more-or-less adequate answer to the most pressing problems of the day. The world now needs to adopt solutions that change the entire system of production and consumption in a fundamental manner, that move societies from conditions of energy and materials surplus to scarcity, and that develop the foresight needed to recognize still-hidden threats to sustainability. This goes far beyond the realm of technical adaptations, and instead requires large-scale social, economic, and political engineering—in an effort to create the foundations for a more sustainable human civilization.

Emerging Issues

CHAPTER 2

Energy, Credit, and the End of Growth

Nathan John Hagens

Human cultures tell stories over time that come to be believed as truths. A prominent example, "Living the American Dream," has implied that the United States is special—that Americans' intelligence and creativity, combined with hard work, initiative, and democracy, largely explain how the country became the world's leading economy and its citizens enjoyed consumption levels that were among the highest anywhere. This narrative still serves as a beacon to people worldwide who aspire to "live the good life."

And despite weak (or negative) growth for the past decade, the U.S. Congressional Budget Office continues to forecast that the U.S. economy will grow 3 percent annually for the next 10 years and beyond, as if it were a natural law. The conventional wisdom is that it is only a matter of time before American ingenuity, technology, and "animal spirits" (a term that economist John Maynard Keynes used to describe the human emotion that drives consumer confidence) will restore the growth trajectories and living standards for which Americans are destined. Key policies and institutions, both in the United States and around the globe, are built on these expectations.[1]

Yet reality has started to diverge from this cultural narrative. Although nominal statistics, such as gross domestic product (GDP) and stock market indices, still broadly signal that everything is fine, the underlying fundamentals paint a different picture. For 95 percent of Americans, real salaries and wages are lower now than they were in 2002. Over the same period, prices have risen sharply for everyday things like energy (up 59 percent), health care (up 18 percent), and education (up 39 percent). U.S. car ownership and oil consumption peaked in 2005, and miles driven peaked in 2007. Wealth inequality is now higher than at any time since the 1820s. Twenty-eight percent of American families have zero savings, and only 43 percent save enough to cover three months of expenses. Half of all U.S. retirees have less than $25,000 in savings of any kind.[2]

It seems that the U.S. economy and its future prospects are not what

Nathan John Hagens is a former hedge fund manager who teaches human ecology at the University of Minnesota. He cofounded and directs the Bottleneck Foundation, which focuses on the long-run aspirations and potential for human society, and was lead editor of theoildrum.com.

Americans have become accustomed to, and not what the government has been forecasting. Why? Is there a common cause? And can anything be done to reverse this course?

Although analysts blame the U.S. economic malaise on a variety of culprits, viewing the situation through a biophysical lens reveals a primary cause: fossil energy. Fossil fuels underpinned the economic miracle of the last century, yet the increasing costs of extraction, particularly for oil, lie at the root of deteriorating economic fundamentals and the gradual loss of the societal benefits that they once provided. Globally, this reduction in benefits is being masked temporarily by a surge in monetary credit and other financial guarantees, but these have practical limits and are, in turn, creating other risks. In short, the waning of the primary drivers of growth—inexpensive "labor" from fossil carbon and freely available monetary credit—suggests that our expectations for continued global economic growth need to be reexamined. (See also Chapter 3, "The Trouble with Growth.")

Energy as the Foundation of Human Economies

In nature, everything runs on energy. The sun's rays combine with nutrients, water, and carbon dioxide to grow plants in a process known as "primary productivity." Animals eat the plants, other animals eat those animals, and so on up the trophic pyramid, with each stage generating an energy input, an energy payoff, and some waste heat. Humanity and its systems conform just as much to this biophysical process as the rest of nature. We combine energy and natural resources with technology and labor to create real things like tractors, houses, and computers. Although we then rank their values with digital representations of money in its various forms, energy remains the foundation of our human ecosystem.[3]

Our energy development trajectory—from using sources such as biomass and draft animals, to wind and water power, and finally to fossil carbon and electricity—has enabled large increases in per capita economic output. This is because, even after accounting for the energy required to extract and process those fuels, large quantities of fuels are still available for other activities. From 1850 to 2010, world human population grew 5-fold, but world energy use increased 20-fold, and fossil fuel use rose more than 150-fold. (See Figure 2–1.) Eighty percent of the nitrogen in our bodies and half of the protein comes directly from natural gas via fertilizers and food, thanks to the Haber-Bosch process, which converts atmospheric nitrogen to nutritionally available ammonia. Whereas people living two centuries ago were made largely from sunlight, we are made largely from fossil hydrocarbons.[4]

Low-cost fossil energy is the foundation of our profits, high salaries, and inexpensive goods and services. We have leveraged our own puny muscle

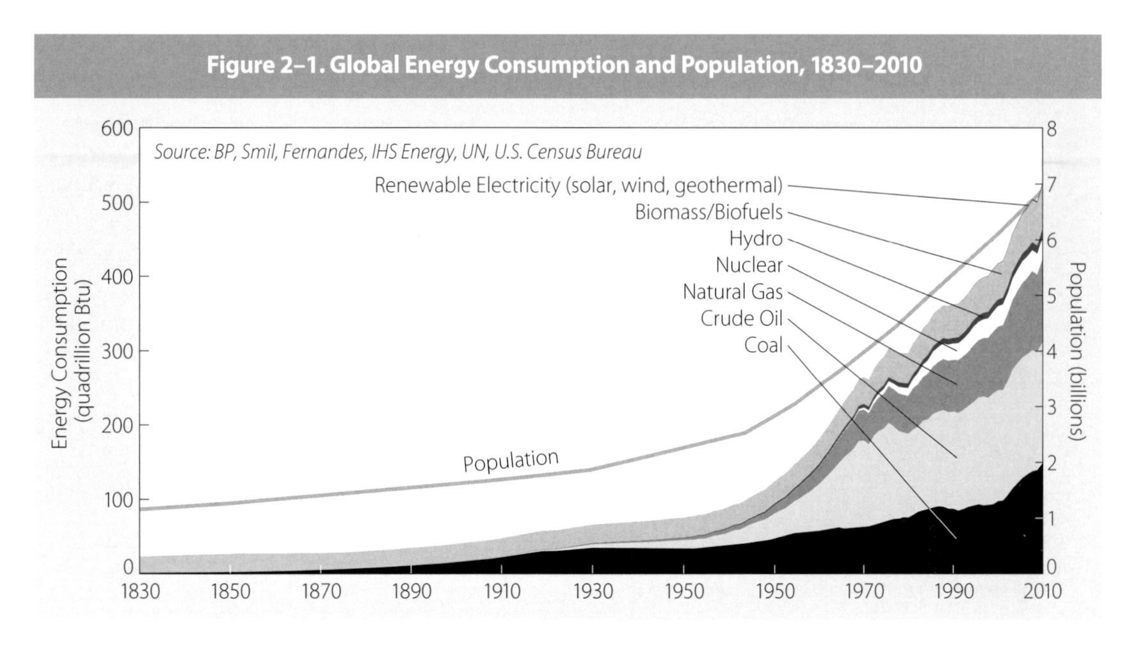

power with the labor of, in effect, billions of far-cheaper fossil energy "slaves." (See Box 2–1.) To the average person, the benefits that we obtain from burning fossil energy may appear as magic: flip a switch and the "slaves" come running—to wash our clothes, cook our meals, heat water, dig building foundations, and perform thousands of other tasks. But the relationship is concrete and straightforward.[5]

Every single good, service, or transaction that contributes to GDP first requires some energy input. Figure 2–2 shows the high correlation of GDP (economic output) to primary energy use, as well as to end-use energy in the form of electricity and transport fuel for ships and trucks. Improvements in efficiency, especially in natural gas plants, complemented energy use as a driver of economic output, but mostly leveled off after the 1990s.[6]

The story of industrialization has been one of applying large amounts of cheap fossil energy to mechanize tasks that humans once performed manually, and to invent many more. We can call this substitution the Big Trade. It was an inefficient trade from the perspective of energy (much more energy was used to accomplish a task), but it was highly profitable from the perspective of human society. Driving a car on a paved road, for example, uses hundreds of times the energy of a human walking, but it moves us 10–20 times faster.

This Trade—replacing human labor with mechanized labor from much-cheaper fossil energy—is largely responsible for the combination of higher wages, higher profits, cheaper goods, and vastly more people that distinguishes the post-industrial world from the rest of human history. And the

Box 2–1. The Power of Fossil Slaves

The average manual laborer expends about 0.6 kilowatt-hours (kWh) of work energy per day—the equivalent of leaving a 100 watt light bulb on for six hours. For thousands of years, that is what civilizational living standards were based on: the combined muscle power (work per unit of time) of groups of human laborers, augmented by a bit of animal muscle power and wind. Most of this was put toward marshaling solar flows and natural systems (soil, forests, rivers, etc.) to generate societal surplus (mostly food).

Then came the Industrial Revolution, starting in the late 1700s, during which we learned how to extract and use the earth's enormous stores of fossil energy. The modern human ecosystem was transformed by this windfall in only a couple of centuries. One barrel of crude oil contains the equivalent of 1,700 kWh of thermal energy, which (at the human average of 0.6 kWh per workday) equals more than 10 years of manual labor. In the United States, a manual laborer averages $29,260 per year, so one barrel of light sweet crude oil represents around $300,000 of manual human labor potential.

Put another way, it costs $260 for an average American to generate 1 kWh of work. The equivalent work can be performed for less than 11 cents with gasoline (at $4 per gallon, or $1.06 per liter) and for about 6 cents with electricity from new coal plants. (See Table 2–1.) Even in poorer countries, the exchange of fossil labor for human labor has been extremely advantageous.

With oil at $60 per barrel (its price in December 2014), the average American has, in effect, almost 6,000 energy slaves to command, replacing what humans used to do. (The average global human commands around 1,300.) Each American uses about 60 barrels of oil equivalent per year in fossil fuels (oil, coal, and natural gas). We are surrounded by millions of unseen, indefatigable fossil carbon slaves.

Table 2–1. Costs of Human Labor versus Fossil "Labor"

Energy Source	Cost per kWh	Multiple of U.S. Laborer	Multiple of Global Average Laborer
	U.S. dollars		
U.S. human laborer	260.00	1	0.22
Global average laborer	57.80	5	1
Typical Bangladeshi laborer	8.26	32	7
Gasoline at $4 per U.S. gallon ($1.06 per liter)	0.109	2,387	530
Electricity from new coal-fired power plant*	0.06	4,336	964
Natural gas at $4 per million cubic feet ($4 per 0.028 million cubic meters)	0.014	18,582	4,129

** Excluding grid costs, i.e., construction, transmission and distribution losses, operations and maintenance.*
Source: See endnote 5.

Figure 2-2. World Primary and Useful Energy Consumption versus GDP, 1980–2008

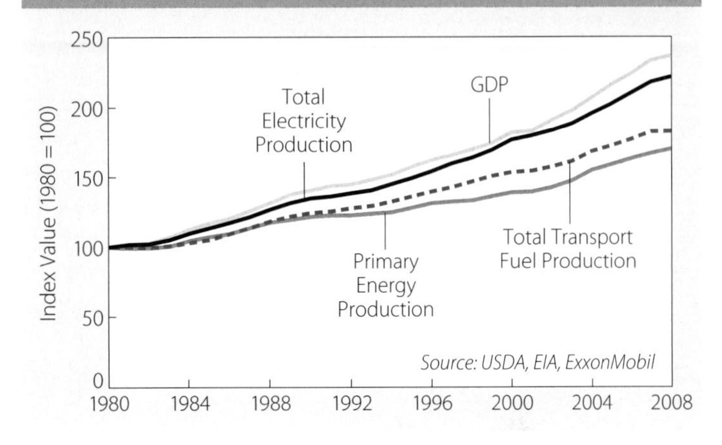

United States has expended more of this fossil magic than any other country—a subsidy that has been part and parcel of the American Dream. Technology has acted as a vector for increasing overall energy use in the economy, adding energy to tasks that humans previously did manually (plowing, driving, manufacturing) and creating myriad new energy-hungry gadgets (microwaves, iPhones, etc.). Globally, each unit of human labor is supported and leveraged by over 90 units of fossil labor. In the industrialized world, it is four to five times that much. Yet most analysts on Wall Street take for granted this exchanging of fossil/mechanical labor for human work.

The Fragility of the Trade

Despite the power in this exchange, its benefits in terms of wages, profits, and cheap goods can be unwound easily if energy prices increase. This is because one unit of human labor is not replaced with just one or a few units of fossil labor, but with hundreds or thousands of units for each task, making the system very sensitive to small increases in energy input costs.

Consider milking a cow using three methods: manual (no energy other than the human labor), semi-automated electric milking machines (1,100 kWh per cow per year, or cow-year), and fully automated milking (3,000 kWh per cow-year). The manual milker, working alone, requires 120 hours of human labor per cow-year, but the semi-automated machines require only 27 hours of labor, and full automation only 12 hours. Let's assume that the human milker is paid $5 an hour working alone. Using electric milkers that run on electricity at $0.05 per kWh, output rises significantly *and*—because cheap electricity substitutes for so many human hours of labor—the wages for the milker increase to $18 per hour with semi-automated milkers and to $33 per hour with the fully automated technologies. This same principle can be extrapolated to many or most modern industrial processes: we save human labor and time by adding large amounts of cheap fossil labor.

In other words, workers are paid according to their productivity, and using lots of cheap fossil energy raises productivity. If that fossil energy gets more expensive, workers' productivity is not raised as much and their labor is worth less. For example, if electricity prices rise to $0.15 per kWh, the manual milker (still) makes $5 per hour, but the semi-automated milking wage declines from $18 to $14 and the fully automated (energy-intensive) wage plunges from $33 to $8 per hour.[7]

The key point is that as the price of energy doubles or triples, the economic benefits from the Trade recede quickly. (This is what Peak Oil is really about; see Box 2–2.) This is especially true for air travel, aluminum smelting, cement manufacturing, and other energy-intensive processes. The reduction in wages that ensues from large energy price increases can be offset only partially by greater efficiency or lean manufacturing measures, because the whole Trade is predicated on large amounts of very cheap energy. This phenomenon of "reduced benefits" is occurring now around the world.[8]

Box 2–2. Peak Oil, or Peak Benefits?

In recent years, much attention has been given to Peak Oil, the notion that basins, regions, countries, and the world as a whole would reach a maximum of oil production and then permanently decline. Since oil is a finite natural resource, the idea that global oil production will one day peak is not a theory but a certainty. But the hyperfocus on the date and production level of such a peak has missed the larger, more relevant point to society: the unfolding decline in benefits that society receives as a result of more-costly extraction.

Peak Oil has never been about running out of oil; it is about quantity, price, and benefits. If global credit markets and economies continue to be stable or grow, we gradually will require more diesel, natural gas, water, and materials to access and produce the next tiers of remaining hydrocarbons. This higher cost will be passed on to consumers and governments, and because societies use so many units of oil for every modern economic transaction, the wages, profits, and cheap goods that we enjoyed when oil was at its historical trend average of $20–30 per barrel are now a phenomenon of the prior era. Peak Oil thus more aptly would have been named "Peak Benefits," and we passed it about 10 years ago.

If fossil energy is undervalued in our economies, then substitutable physical labor is, by the same token, overvalued. A truck loaded with 3,000 pounds (1,361 kilograms) of goods can move them 30 miles (48 kilometers) in an hour or so for roughly 1.5 gallons (5.7 liters) of gasoline, at a U.S. cost (at this writing) of less than $5. It would take eight strong men (at 1.5 miles, or 2.4 kilometers, per hour if they were lucky) 20 hours to push the same weight over the same distance. Even at the U.S. federal minimum wage (currently $7.25 per hour), this job would cost $1,160! The $5 paid for gasoline would have afforded the laborers only $0.031 (3.1 cents) per hour each to do the same work. This partially explains why real wages peaked in the United States in 1973, but productivity (and profit/wealth creation) continued to rise. Over time, given such an enormous cost differential, the global owners of capital chose to hire fossil slaves instead of real human labor.

The Importance of Energy Return on Energy Invested

It takes energy to get energy. For example, it requires about 245 kilojoules of energy to lift 5 kilograms of oil 5 kilometers out of the ground. For decades, at least until recently, oil was relatively inexpensive for the simple reason that it took minimal energy to extract it from the ground. However, technologies (such as horizontal drilling and shale fracturing) that are needed to extract oil in many "new" discoveries are increasingly expensive in terms of energy.

This matters in the human economic system because we need to use an increasing share of the energy we produce in order to generate the energy we need. The amount of energy available to spend on civilization—to create and support roads, symphony orchestras, iPods, haute cuisine, hospitals, etc.—is only the amount left over after other energy and resources have been expended to harvest and distribute that energy. As marginal energy supplies require increasing amounts of energy to harness them, society has less surplus energy to expend on other pursuits. While the media and most government statistics report gross energy, it is the net energy that really matters.

One statistic that measures this biophysical situation is the energy return on energy invested (EROEI). The extraction of finite resources typically follows a "best first" pattern. In the case of oil, we first tapped surface seeps, then in the industrial era we learned to use seismic surveys to reveal deposits far underground. Now we are exploring deep-water and subsalt reservoirs, and most recently have been hydro-fracturing "tight oil" formations. (See Box 2–3.) At each phase, the EROEI of discovering oil has fallen, declining from more than 100:1 to less than 10:1 since the beginning of the modern oil era.[9]

Since 2002, oil production costs (in money terms) have increased 17 percent a year while inflation has averaged only 2 percent per year. This contributes to a vicious cycle, as all the inputs to energy extraction are affected by increases in fuel costs. In 2001, major oil companies needed a price of $9 per barrel of oil for their revenues to cover their costs, dividends, and capital expenditures. Today, that figure is $120 a barrel. At $90 and below, many of the new oil plays become uneconomic to drill.[10]

Under current trends, an ever-increasing share of total GDP will be allocated to the energy sector. In 2013, fully a third of capital expenditures for companies listed on the Standard and Poor's (S&P) 500 stock market index were in the energy sector. Since December 2007, or roughly the beginning of the great recession, shale oil states (Colorado, North Dakota, Pennsylvania, Texas, and West Virginia) have added 1.4 million jobs, while non-shale states have lost 424,000 jobs. If our objective remains to just increase GDP,

Box 2–3. A Brief Guide to Fracking

In 2005, it was widely believed that the United States was on the cusp of a natural gas crisis because existing natural gas wells were depleting faster than new wells were being brought on line. One industry publication forecast that by 2025, there would be a U.S. natural gas shortfall equal to nearly 30 percent of demand. In 2003, the late Matt Simmons, author of *Twilight in the Desert*, projected an even earlier crisis, affirming with "certainty" that by 2005 the United States would enter a long-term natural gas crisis for which the only solution was "to pray."

As it turned out, the evolution of two well-established technologies sharply changed the near-term trajectory of U.S. oil and gas production. The first, a drilling technique called hydraulic fracturing ("fracking"), was developed in the late 1940s to promote higher production rates from oil and gas wells. Fracking involves pumping water, chemicals, and a proppant down an oil or gas well under high pressure to break open channels (fractures) in the reservoir rock, allowing the trapped oil and gas to flow to the well bore. The proppant, most commonly sand, is used to hold the channels open.

The second technique, horizontal drilling, was invented decades ago and has been used widely in the oil and gas industry since the 1980s. It involves drilling down to an oil or gas deposit and then turning the drill parallel to the formation to drill along its length and thus access a greater fraction of the deposit. These "laterals" can be over 3,000 meters long.

In the late 1990s, an engineer-businessman named George P. Mitchell paired fracking with horizontal drilling in the Barnett Shale formation in Texas. Prior to Mitchell's experiments, the oil and gas trapped in these shales had been too expensive to produce, but use of the combined techniques altered the economics, and by 2005 a fracking revolution was driving a shale oil and gas boom. In 2006, U.S. natural gas production reversed course, and it has risen every year since. In 2011, it exceeded its previous all-time high mark, set in 1973. By 2013, U.S. gas production was 33 percent above the 2005 level, and the United States was the largest natural gas producer in the world.

These new techniques also were enormously successful in the oil industry. In 2009, U.S. oil production reversed nearly 40 years of decline and began rising at the fastest rate in U.S. history. By 2013, U.S. production had increased by 3.2 million barrels per day from 2008 levels—accounting for an astounding 83.6 percent of the total global increase in oil production.

The fracking revolution is credited with providing huge economic benefits to the United States. A report by the McKinsey Global Institute concluded that shale gas and oil will add $380–690 billion per year to U.S. GDP and create 1.7 million permanent jobs in the process. But some have characterized fracking as a mirage, a flash in the pan, a Ponzi scheme, and an environmental nightmare that threatens our water supplies, causes earthquakes, and has the potential to derail a clean energy revolution.

Environmental issues began to surface as fracking moved into populated areas unaccustomed to oil and gas development. People near fracked wells cited various maladies, and cases of contaminated water were blamed on fracking. Proponents insisted that such contamination was impossible because thousands of meters of rock lay between oil and gas deposits and water reservoirs. In September 2014, a two-year study funded by the National Science Foundation concluded that reported contamination was due to well leaks near a water source, and not to the migration of fracking fluids into water reservoirs. Yet fracking enabled the economic production of the well and, ultimately, the contamination of the water.

Several studies have pointed to a direct link

Box 2–3. continued

between increased fracking and the rise in earthquakes in an area. The suspected cause is not fracking itself, but rather the process of pumping the fracking wastewater down disposal wells, which may decrease the friction along a fault and enable an earthquake to take place. Prior to the fracking boom, Oklahoma averaged one earthquake a year with a magnitude of at least 3 on the Richter scale. During the first half of 2014, Oklahoma had 258 earthquakes in that range, nearly twice as many as California.

Because production in many fracked wells declines rapidly—by as much as 90 percent of initial extraction rates in the first two years of operation—more and more wells must be drilled just to maintain production. Thus far, the inventory of drilling sites has been sufficient to allow for increased production, but many critics argue that the shale boom is really a bubble, and that once production begins to decline, it will do so rapidly. Further, companies that are borrowing money to invest heavily in shale oil and gas production may find themselves unable to pay back that money once overall production begins to decline.

Fracking perfectly illustrates the fact that energy production always comes at some cost. While providing economic benefits, there are also environmental costs. But those suffering the environmental impacts are not necessarily the same as those enjoying the economic benefits. The future of fracking will hinge in part on whether and how this discrepancy is resolved.

—*Robert Rapier, author and chief investment strategist,* Investing Daily's *Energy Strategist* *service*

Source: See endnote 10.

we can keep growing gross energy production by locating and exploiting deeper and deeper pockets of fossil hydrocarbons.[11]

But eventually, the entire economy would have to be devoted to supporting a giant mining operation—meaning that there would be little spending left for art, education, medicine, or any other sector. In addition, the higher costs to access this lower-quality energy would result in far fewer benefits for society.

In effect, a declining EROEI acts as a tax on the rest of society, especially one built on expectations of continued high EROEI. Media outlets tend to overlook this decline in net energy and the reduced benefits, focusing instead on the new surge in gross U.S. "oil production." (Those figures include a large share of natural gas liquids—not really oil but a byproduct of drilling for gas—and billions of liters of corn ethanol, which is not an energy source but a conversion of soil, natural gas, and corn into liquid fuel; the EROEI is barely over 1:1.) The media rarely note that capital expenditure requirements are rising faster than oil prices, or that exploiting shale formations requires an enormous increase in diesel use, or that the resulting oil has a higher API gravity (meaning, actually, that it is lighter), which exaggerates the energy content per barrel by 3.5–10.7 percent.[12]

Despite having "plenty of energy," higher physical costs suggest that energy likely will rise from a historical average of 5 percent of GDP, to 10–15 percent of GDP or higher. (Because it is not just gasoline prices and home utility bills that matter, but the entire embedded infrastructure on which the global supply web depends. Energy, particularly oil price increases, ripples through every aspect of our lives.) In the short run, however, we can paper over these physical cost increases, well, with paper (money).[13]

Debt and Energy

Money routinely is mistaken for wealth, but money and financial instruments are simply markers for the four kinds of real capital: natural (oil, trees, rivers), built (houses, tractors, computers), social (relationships, networks), and human (health, skills, knowledge). Money is essentially a claim on a certain amount of energy. When the U.S. economy began a period of explosive growth in the early 1900s, money, rather than energy or resources, was the limiting factor. There was so much wealth in natural resources that the country needed ways to turbocharge the broader economy so that anyone with skill, good ideas, or ambition could undertake productive ventures. It was around this time that the world's central banks created rules for commercial banks, with the intent to increase the flow of money to match the productive output of industrial economies. Creditworthy individuals and businesses could now obtain loans from commercial banks, which were required to keep a small portion of their assets on reserve with a central bank.[14]

Business schools teach that credit creation is a series of consecutive bank "intermediations," where an initial deposit of wealth ripples through the banking system and, by being repeated many times since only a fraction is required as reserves, creates additional money. But this is true for only about 5 percent of money coming into existence. The reality for 95 percent of money creation is profoundly different. If a businesswoman needs $100,000 to start a car wash and her local commercial bank deems her creditworthy, $100,000 is entered electronically into her checking account (an asset for her and a liability for the bank), and at the same time a receivable (or IOU) from the businesswoman is entered on the bank's books (a liability for her and an asset for the bank).[15]

But something extraordinary has just happened. The act of lending normally means a transfer of an existing commodity to its exclusive use somewhere else; if someone lends a hammer to his neighbor, he loses the use of it until the neighbor gives it back. However, even though the assets and liabilities are in balance, this new credit extended by the bank *does not remove $100,000 of purchasing power from anywhere else in the system.* Banks lend not when they have new deposits, but when they have demand for loans from creditworthy customers. In effect, banks do not lend money, they create it. Were it only so easy to create real wealth, i.e., energy![16]

Of all the money outstanding in the United States (around $60 trillion), only about $1 trillion is physical currency (cash). The rest can be considered corporate, household, or government debt. In financial textbooks, debt is an economically neutral concept, neither bad nor good, but just an exchange of time preference between two parties who want to consume at different times. However, if cash is a claim on energy and resources, debt is a claim on *future* energy and resources, and several things happen when debt is issued that have much different impacts than the textbooks claim.

First, all the while that debt is being issued, the highest-energy-gain fuels are depleted, making energy (and therefore other things) generally more expensive for the creditor in the future than for the debtor in the present. In such a situation, people who choose to save are "outcompeted" by people who choose to consume by taking on debt. At some point in the future, at least some creditors will get less than they are owed, or even nothing.

Second, increasingly more credit must be issued to mask the declining benefits of the Trade, lest aggregate demand plunge and the ranks of the poor swell.

Third, just as lower EROEI means that the productivity of energy extraction is declining, so too does the productivity of debt decline, as we have to add more and more debt to get small increases in GDP. When we add $1 trillion in debt and our output goes up by at least $1 trillion, there is no problem. But if we continually generate less and less GDP for each additional debt dollar, we are in unsustainable territory. This can be measured by "debt productivity," or how much GDP is generated for an additional dollar of debt (the ratio of GDP growth to debt growth).

Blek

Motorcycle taxi driver counting his earnings in Jakarta, Indonesia.

Since 2008, the Group of Seven nations (Canada, France, Germany, Italy, Japan, the United Kingdom, and the United States) have added about $1 trillion per year in nominal GDP, but only by increasing debt by $18 trillion-plus. This has significantly lowered their debt productivity, and when this ratio gets low (or approaches zero, as is the case now), new debt basically is just an exchange of wealth for income. With money as well as energy, these countries are witnessing a "Red Queen phenomenon"—running harder to stay in the same place—as they add increasingly large debt burdens in order to keep consumption up and GDP (slowly) growing. Looking only at GDP and stock prices, everything seems fine; but looking at energy extraction costs and new and growing claims on future energy and resources, the picture looks considerably more ominous.

From an ecological perspective, all of these existing debts are claims on the energy and natural resources required to repay them (with interest)—energy and resources that have yet to be extracted. In the past decade, the global credit market has grown 12 percent per year, yet GDP has grown only 3.5 percent annually, and global crude oil production has grown less than 1 percent annually. And, since 2008, despite energy's fundamental role in economic growth, it is access to credit that has been supporting economies. As long as interest rates (government borrowing costs) are low and market participants accept the debt, this can go on for a long time, although energy costs are likely to continue rising and the benefits from the Trade to continue declining, creating other societal pressures. The government takeover of the credit mechanism seems unlikely to stop soon, but if it does, both oil production and oil prices will be considerably lower. (This is not necessarily a good thing for long-term economic health; see Box 2–4.)[17]

Debt temporarily makes gross energy feel like net energy as a larger amount of energy is burned despite higher prices, lower wages, and lower profits. Increasing gross energy also adds to GDP. But over time, as debt increases gross energy and as net energy stays constant or declines, a larger portion of our economy becomes involved in the energy sector. At some point in the future, important processes and aspects of non-energy infrastructure will become too expensive to continue (think of the fully automated milking example). Even more worrying is that, faced with higher costs, energy companies increasingly are following the societal choice to use debt to pull production forward in time. In this environment, we can expect total capital expenditures to keep pace with total revenue every year, while net cash flow becomes negative as debt rises. Not many companies can afford to lose money every day and stay in business. In a world awash in debt, Peak Oil may well be evidenced by energy company bankruptcies as opposed to higher prices.

Governments (and corporations) are now facilitating an increase in gross energy even as net energy benefits (wages, affordability of goods, etc.) have peaked and are declining. At the same time, debt expansion reflects people's increasing claims on what they believe they own and have access to in the future. In the past few years, central banks have subsidized our consumptive lifestyle to the tune of more than $14 trillion in such measures, allowing energy extraction to continue apace by temporarily obscuring EROEI effects and signals. And our situation does not entail only government debt; all of our financial claims are debt relative to the available natural resources.

Box 2-4. Oil Prices: Walking on a Wire

Crude oil prices fell by over $50 per barrel from their 2013 high of $111 to their December 2014 price of about $60 per barrel. Given the enormous benefits that oil provides society, one might think this a good thing—and it might be, if 1) extraction costs for energy companies (in both monetary and energy terms) were declining as well, and 2) the financial health of energy companies was sound and improving.

Neither is true. Typical costs of extracting U.S. shale oil are $60–$80 per barrel. Although consumers love low gas prices, no oil company CEO (or shareholder) is happy with market prices more or less equal to extraction costs. Outside of the Organization of the Petroleum Exporting Countries (OPEC), those costs have increased 17 percent per year since 2002 as oil firms use more expensive technology, fracking techniques, and deeper wells. Below $85 per barrel, many companies are pulling drilling rigs and waiting for higher prices to drill new wells.

National oil companies like those in Saudi Arabia, Russia, and Iran, which are run by their countries' respective governments, require selling prices of over $100 per barrel to avoid budgetary shortfalls and cuts in social programs. Energy companies, especially those drilling the shale plays, have taken on enormous debt burdens to fund the high capital expenditures necessary for complicated drilling. They require high oil prices in order to generate profits to pay interest and principal back to banks.

Finally, the financial representations of physical things (e.g., futures and options on crude oil) can end up affecting the physical thing itself. In late 2014, the average price of a barrel of oil plunged from $95 in August to $60 in December and was expected to drop further, at least in the United States. Worries about rising energy costs were the furthest thing from most minds at that time; however, the long-term price trend is upward not downward. In this era, low oil prices are a symptom of our declining energy surplus: our ability to afford oil is declining faster than the aggregate depletion rate of oil fields.

But when market prices fall below the cost of production, the result is less drilling of new wells and less-stable capital structures for what is arguably the most important industry in the world. Already in November 2014, new oil drilling permits fell by 40 percent. Given that many wells (such as those tapping Bakken shale) deplete at rates of 80–90 percent in the first two years, low oil prices, while at least temporarily beneficial to consumers, are the seeds of destruction for many oil companies, which means the next wave of higher prices at the pump.

Source: See endnote 17.

To return to an earlier point: energy, not money, is what we really have to spend or save.[18]

All of this leads to a scenario that is unpleasant but easily imaginable. Higher energy costs over time, particularly for oil, will have ripple effects through any economy that has built itself on large energy input requirements (such as those of the entire industrialized world). The first two likely casualties will be 1) highly energy-intensive industries and practices, which will gradually become uneconomic (including and notably in the energy sector itself), and 2) everyone who will experience the impact of widening and deepening poverty. Everything we do will become more expensive (or less affordable) if we cannot reduce the energy consumption of specific processes faster than extraction costs rise.

Conclusion

The summary of our situation is both simple and challenging:

- Energy underpins our society, and cheap energy underpins our high living standards, policies, and expectations.
- The production of energy requires, first and foremost, energy.
- The energy return on fossil fuels is declining because we have accessed the easy and somewhat-easy stores.
- We have offset the rising extraction costs with something that we could do easily: increase debt.
- High market prices (and eventually even higher prices) are required if production levels are to be maintained.
- However, these high prices destroy demand, provide fewer benefits to society, and lead to recession, excess debt-based claims on future production, and more-severe social inequality.

The issue we face is not a lack of fossil energy nor viable renewable energy technology, but a vast built infrastructure, complex supply chains, and a socioeconomic system that requires ongoing growth each year in order to service prior financial claims. Such a system requires 4–6 cent per kWh electricity and $20–$30 a barrel oil. Renewable power and efficiency gains are important, but they cannot overcome the eventual unraveling of benefit expectations and, ultimately, of the financial claims built into the current system. Thus, we need to combine planning for a low-carbon future with preparing for a lower-consumption future, which includes renewable energy (but not with an expectation of continuing present consumption levels). If those who care about the environment fail to integrate this into their thinking, they risk becoming largely irrelevant in the coming years as the economy, wages, and job prospects deteriorate.

Andrew Ballantyne

Seismic vibrator trucks being used to explore for oil in the Egyptian desert.

We have two major, interrelated problems: one physical, one social. The first is that we are hitting limits to growth thresholds: energy costs, energy use per capita, financial marker overshoot, water shortages, greenhouse gas emissions affecting the biosphere and oceans, biodiversity loss, etc. But the social aspect that compounds those problems is that modern democracies struggle to acknowledge or even understand these risks. "The end of growth" and "Energy is what we have to spend, money is just a marker" are not the pithy phrases likely to secure a politician re-election or popularity. Democracies will respond to these long-term paradigm shifts only if more people understand the true nature of our problem, and do not blame scapegoats for declining wages, salaries, and social mobility. The true villain is our fossil slaves, which are now asking for unprecedented—and unsustainable—pay raises. The way of life that they originally enabled is increasingly beyond our means.

We have entered a period of unknown duration where things are going to be tough. But humanity in the past has responded in creative, unexpected ways with new inventions and aspirations. We tend to crave one solution, while in reality for what we face there is no such thing. Many approaches advocate the means (the way things are produced) yet not the ends (growth, resource consumption), while others aim to alter the ends (GDP) without addressing the means (the way things are produced/delivered). Some policy choices, including banking reform, a carbon or consumption tax, and moving away from GDP as a proxy for well-being, are good long-term ideas. But our declining energy-quality situation will act increasingly as a tax on our societal surplus, such that enacting any of these "more sustainable" options runs the risk of tipping the global economy into depression or worse.

We urgently need institutions and populations to begin to prepare, physically and psychologically, for a world with the same or less each year instead of more—a mindset that is not in our collective psyches or even imaginations. Millions of small "solutions" are needed to put humanity on a better course, some of which will play tiny roles, and some significant. Among the needed action steps are a change in values by those who no longer have access to the smorgasbord of benefits of the last generation; community-driven initiatives that fill gaps left by the loss of government funding; and reducing dependency on those goods, services, and processes that will undergo large cost increases. Energy and environmental education also are necessary, so that more people understand why our situation is not the fault of some political party or group of people, but a natural result of higher costs for our most important input. Ultimately, we face not a shortage of energy, but a longage of expectations.

CHAPTER 3

The Trouble with Growth

Peter A. Victor and Tim Jackson

In July 2013, a remarkable conference took place in a meeting hall of the French National Assembly in Paris. Current and former government ministers from France, Sweden, Greece, Spain, and Brazil, under the aegis of the president of France, François Hollande, met to explore nothing less than modern economic heresy: the abandonment of governments' longstanding commitment to continuous economic growth—and its replacement, at least in the view of some attendees, with goals focusing on well-being, equality, and environmental health.[1]

The conference was remarkable not only for the number of officials willing to think outside the conventional economic box, but because the topic barely registers on the radars of officials outside Europe, and is even less known to the general public. Indeed, the need for economic growth continues to be unquestioned dogma in most of the world, even in governments that claim to be striving for sustainability.

Concern over the ongoing expansion of the world's economies is driven largely by the unsustainable burden that this relentless growth imposes on the planet's life-support systems. Evidence of this burden has been accumulating for decades. For example:

- Five assessment reports issued by the Intergovernmental Panel on Climate Change between 1990 and 2014 document, with steadily increasing certainty, the growing human influence on Earth's climate.
- The 2005 *Millennium Ecosystem Assessment* concluded that roughly 60 percent of the services provided by nature to humans are in decline.
- Work since 2009 on "planetary boundaries" has identified factors that drive nine major environmental phenomena—including climate change, biodiversity loss, and nitrogen pollution—and suggests that in several cases the boundaries have already been crossed.
- The 2014 *Living Planet Report* documents that populations of vertebrate species have declined by half since 1970.[2]

Peter A. Victor is a professor of environmental studies at York University. **Tim Jackson** is a professor of sustainable development at the University of Surrey and also an ESRC professorial fellow on Prosperity and Sustainability in the Green Economy (PASSAGE).

These and other compelling studies pose a strong challenge to modern ideas of progress, and they suggest that a commitment to economic growth is a hidden threat to sustainability. Fortunately, humans have millennia-worth of experience building economies that are not driven by a growth imperative. And today, research suggests that modern economies could provide jobs and reduce inequality (perhaps more effectively than today's economies do)—even as they lighten humanity's impact on the environment—without pursuing economic growth. Given the need for economies designed to grow slowly or not at all, especially in physical terms, the question is whether policy makers and the public worldwide can summon the courage and open-mindedness to forgo economic growth as a policy priority.

Economic Growth as a Policy Objective

Anyone born after the middle of the twentieth century can be excused for thinking that economic growth has always been a top priority for governments. But as Heinz W. Arndt observed in his history of economic growth, "There is in fact hardly a trace of interest in economic growth as a policy objective in the official or professional literature of western countries before 1950." This will come as quite a surprise to those accustomed to the endless stream of statements by politicians, pundits, the media, and economists about the importance of economic growth.[3]

Economic growth as a policy objective emerged after World War II as an effort by governments to achieve full employment for their citizens. In the 1930s, British economist John Maynard Keynes had argued persuasively that while no mechanism exists in the private sector of capitalist economies to guarantee full employment, government spending could be used to prime the economic pump and stimulate job creation. His theoretical arguments were borne out by the experience of World War II, during which government spending increased dramatically, especially among the Allied nations, and unemployment largely disappeared.[4]

During and following the war, with the bread lines of the Great Depression fresh in their memories, governments in many countries adopted full employment as an explicit policy objective, believing that Keynes had equipped them with the means to achieve it. Full employment required that total expenditures rise continually to pay for the new infrastructure, factories, and equipment that made full employment possible. So governments began to pursue economic growth as a means of achieving their full employment goals. Within a few years, likely because of the Cold War and the global arms race, economic growth became an objective in its own right.

In 1960, member countries of the newly established Organisation for Economic Co-operation and Development (OECD) declared in the organization's charter that: "The aims of the OECD shall be to promote policies designed to achieve the highest sustainable economic growth and employment and a rising standard of living in Member countries." From then on, economic growth has been among the top economic policy objectives of governments, not only in OECD member countries, but in international organizations and countries around the world. (See Box 3–1.)[5]

Box 3–1. What Is Economic Growth?

Economic growth refers to an increase in the goods and services produced by an economy during a given period, as measured by the rate of change in gross domestic product (GDP), excluding inflation. In its simplest terms, GDP is a measure of economic activity—or "busyness"—in an economy.

When GDP is divided by population, the result is GDP per capita, which is often used to measure the "standard of living" of a country. Some countries, such as Luxembourg and Singapore, have very high GDP per capita, even though their total GDPs are low. China and India have large GDPs, but relatively low GDP per capita. GDP and GDP per capita are snapshot measures of an economy. Economic growth is about their change over time.

The benefits of economic growth are distributed unequally within and among countries. The same is true of the costs of economic growth. For example, climate change, caused mainly by the emission of greenhouse gases when energy is used to power economic growth, is just one of many examples of this maldistribution of costs: those most vulnerable to the impacts of climate change have been the least responsible for causing it.

The important point is that belief in the indispensability of economic growth, while deeply rooted in governments virtually worldwide, is quite recent. The common view that growth has always been an important objective of government is mistaken. That growth is inextricably bound up with human nature is an even greater mistake, if it makes us think that there really is no alternative to economic growth. Understanding that growth is not a necessary goal of government policy is critical if we are to imagine alternative economic futures.

The Negative Consequences of Economic Growth

While economic growth has brought higher living standards and jobs for many people, along with tax revenues for governments, it has been achieved at the cost of depleted soils and aquifers; degraded lands and forests; contaminated rivers, seas, and oceans; disrupted cycles of carbon, nitrogen, and phosphorous; and more. In short, economic growth is not an unqualified good. And these environmental costs, along with the social costs of unequal growth, can be substantial.[6]

In the 1970s, economist Herman Daly considered the possibility that economic growth can have such serious negative consequences that they outweigh the benefits of growth. When growth does more harm than good, he explained, it should be described as "uneconomic," since it results from an uneconomical use of resources. Daly and his colleagues developed the Index of Sustainable Economic Welfare (ISEW) to capture the good and bad of economic growth and to give a more accurate measure of economic advance. The ISEW subtracts from GDP the value of unwanted side effects of economic activity—such as the costs of commuting; "defensive" private expenditures on health; pollution; and the depletion of natural resources—and adds in the value of activity that advances well-being and is overlooked by GDP, such as unpaid household work.[7]

Nick Normal

Used windows available for reuse at a nonprofit center in Queens, New York.

Daly's team concluded that in the United States from 1950 to 1990, ISEW per capita increased far more slowly than GDP per capita—well-being lagged far behind output—and that in the final decade (1980–90), ISEW per capita actually declined. Uneconomic growth had arrived in the United States. Similar studies for other countries and regions using the ISEW and its sister measure, the Genuine Progress Indicator, have produced similar results.[8]

In view of the increasingly mixed record of economic growth, Daly concluded that an alternative to growth economies was needed. He advocated for a "steady-state" economy in which the materials and energy used to produce goods and services is kept roughly constant (through recycling, substitution of services for goods, and other materials-saving strategies). Daly distinguished between growth and development, arguing that economies could and should continue to develop indefinitely, but without growth of the economy's material requirements.[9]

Defenders of economic growth often assert that slow or no growth will result in mass unemployment and misery, that the best way to reduce the costs of economic growth is more growth, and that prices and technology will ensure that economic growth is sustainable over the long run. Even many advocates of sustainable economies argue that growth is necessary. The Global Commission on Climate and Economy, led by Sir Nicholas Stern, launched its recent *New Climate Economy* report under a bold headline in favor of "better climate, better growth." Meanwhile, the OECD's annual

Going for Growth report continues to advise member countries on how to increase their rates of economic growth, even though other reports from the OECD propose "green growth" and "inclusive green growth," adding a social justice dimension.[10]

Thus, the critique of growth, which has a long and serious pedigree, has generated serious pushback. Who is correct—the critics of growth or its defenders? Central to the debate is the answer to these questions: Can economic growth be designed in a way that reduces its "uneconomic" costs, even as growth continues indefinitely? Or must growth be abandoned in order to put the world's economies on a sustainable path?

Decoupling Economic Growth from Throughput

The environmental costs of economic growth come from the increasing use of "throughput": the materials (i.e., biomass, construction materials, metals, minerals, and fossil fuels) used to support economic growth. Obtaining increasing supplies of these materials has led to deforestation, the degradation and loss of soil, the removal of massive quantities of material to access underground resources, transformation of the landscape, and more-frequent and more-serious pollution as ever more remote sources of materials, especially fossil fuels, are accessed. Most of these materials remain within the economy for a very short time: fuels for only moments upon use, and many other materials, even with recycling, for less than a year—although some remain much longer, such as building materials and precious metals.[11]

After use, these discarded materials and the dissipated energy are disposed of back into the environment, which has a limited capacity to absorb them. When this capacity is exceeded, a wide range of environmental problems arise. In the early days of industrialization, these problems were primarily local (e.g., polluted rivers and urban air, municipal waste dumps, mine tailings), but with global economic expansion, the associated environmental problems became regional (e.g., acid rain, hazardous waste shipment and disposal) and now global (e.g., acidification of the oceans, loss of biodiversity, climate change).

A critical question is whether throughput, especially those components that do the most damage, can be decoupled from economic growth. If it can, then at least the environmental reasons for questioning the sustainability of economic growth can be addressed. Some analysts are very optimistic about the potential for decoupling growth from throughput, by shifting consumption from goods to services, better design of products and processes, recycling more, substituting scarce materials with more abundant ones, and replacing fossil fuel energy with energy from renewable sources.[12]

Significant improvements in efficiency are possible, especially if determined efforts are made to achieve them. It is by no means certain, however, that these actions will be sufficient to meet wide-ranging economic, social, and environmental objectives, especially in the long term. Ernst von Weizsäcker and his colleagues speak of a "factor 10" economy in which the throughput requirements per dollar of GDP are reduced by a factor of 10. Such a reduction would allow economies to increase their GDP 10 times without any increase in throughput. Alternatively, they could reduce their throughput by 50 percent if GDP increased only five times.[13]

This tradeoff between GDP growth and throughput reduction has two critical implications. First, the greater the rate of economic growth, the faster must be the decline in the rate of throughput (i.e., throughput per unit of GDP) to achieve any desired level of total throughput reduction. There already are many indications that the capacity of the biosphere to absorb the wastes generated by the world's economies has been exceeded in several important respects. Therefore, global throughput will have to be reduced as swiftly and as equitably as possible, to bring economic and environmental systems back into some sort of balance. And this is without considering that some components of throughput (e.g., radioactive waste, heavy metals, carbon emissions) accumulate in the biosphere, requiring even greater reductions in throughput to achieve reduction targets. Hence, some decoupling is required even in the absence of economic growth. But even more decoupling is required if economies grow, and the faster they grow the faster must be the rate of decoupling.

The second critical implication of the relationship between rates of economic growth and decoupling is to consider what happens after a substantial, say tenfold, increase in GDP—even assuming a tenfold or more level of decoupling. An economy growing at 3 percent per year will experience a tenfold increase in GDP after 78 years, which is about the average lifetime of a person born in an industrialized country. Throughput per dollar of GDP will have to shrink to 10 percent of its current value over that period to avoid an increase in total throughput, which is very ambitious. After that, if economic growth continues for another human life span without an increase in throughput, throughput per dollar will have to be only 1 percent of what it is today simply to avoid an increase in the total. At some point, this process must come to an end and economic growth must cease if sustainability is to be achieved.

These arithmetic examples can help scope out the extent to which decoupling is required, but they cannot tell us anything about what might be feasible. Fortunately, Vaclav Smil makes a recent attempt to assess feasibility in his book *Making the Modern World: Materials and Dematerialization*. Smil provides a comprehensive, detailed account of decoupling from the first Industrial Revolution to the present day. He makes the important distinction between relative and absolute decoupling, as others have done. Relative decoupling is about reductions in throughput per dollar of GDP, and absolute decoupling occurs when total throughput, or some important component of it, declines while GDP increases. Smil provides plenty of evidence and numerous examples of relative decoupling, and he expects it to continue well into the future. In contrast, he is very skeptical about the prospects for absolute decoupling, and yet this is what those who maintain that growth can continue indefinitely rely on.[14]

Relative decoupling does not automatically lead to absolute decoupling for a number of reasons. The first is the Jevons paradox, an insight put forward by William S. Jevons in his 1865 study of Britain's coal industry, which revealed that improvements in efficiency lead to reductions in operating costs, but that lower operating costs often induce increases in use. It is "a confusion of ideas to suppose that the economical use of fuel is equivalent to diminished consumption. The very contrary is the truth." The Jevons paradox, or "rebound effect" as it is now sometimes called, is a pervasive relationship that explains much of the disconnect between relative and absolute decoupling.[15]

The other two factors that explain the disconnect between efficiency improvements and absolute reductions in throughput are increases in population and increases in the general level of consumption.

João Ramid/ Norsk Hydro ASA

Bauxite ore being stockpiled at a mine in northern Brazil.

Concluding his book, Smil writes, "to stress the key point for the last time, these impressive achievements of relative dematerialization have not translated into any absolute declines of material use on the global scale." He goes on to say that "the global gap between the haves (approximately 1.5 billion people in 2013) and the have-nots (more than 5.5 billion in 2013) remains so large that even if the aspirations of the materially deprived four-fifths of humanity were to reach only a third of the average living standard that now prevails in affluent countries, the world would be looking at the continuation of aggregate material growth for generations to come."[16]

So much for global decoupling. At the national level, Smil observes that, "Clearly, there is no recent evidence of any widespread and substantial dematerialization—be it in absolute or . . . per capita terms—even among the world's richest economies." However, he does point to Germany and the United Kingdom as examples of a few countries where "overall material inputs have stabilized or have even slightly declined . . . while some of their specific inputs continued to rise." Smil acknowledges that this promising result may be due to changes in trade patterns, and this is exactly what has been shown in other research studies. The shift in manufacturing from industrialized to developing countries has entailed a shift in the location of where materials enter the interconnected economies of the global economic system, rather than any real reduction.[17]

Another recent study by Tommy Wiedmann and colleagues traces the material inputs (i.e., biomass, construction minerals, fossil fuels, and metal ores) embedded in the consumption of 186 countries from 1990 to 2008, and makes very clear the connection between international trade and the absence of absolute decoupling. The authors conclude: "As wealth grows, countries tend to reduce their domestic portion of materials extraction through international trade, whereas the overall mass of material consumption generally increases. With every 10% increase in gross domestic product, the average national MF increases by 6%." MF refers to the "material footprint" of nations, which includes all of the materials used to support consumption in countries irrespective of where the materials are obtained.[18]

Wiedmann and colleagues observe that: "The EU-27 [27 member countries of the European Union], the OECD, the United States, Japan, and the United Kingdom have grown economically while keeping DMC [direct material consumption] at bay or even reducing it, leading to large apparent gains in GDP/DMC resource productivity. In all cases, however, the MF has kept pace with increases in GDP and no improvements in resource productivity at all are observed when measured as the GDP/MF." These findings are illustrated in Figure 3–1, where from 1990 to 2008, the material footprint of OECD countries in total moved in step with GDP, while their direct material consumption showed relative decoupling (and absolute decoupling during recessions).[19]

The most reasonable conclusion to draw from studies such as those of Smil and Wiedmann et al. is that there is very little precedent for absolute decoupling and no foundation of experience on which to base a realistic expectation for the degree of decoupling required for sustainability. So while one may speculate boldly about the future prospects for absolute decoupling of throughput from economic growth, and thus maintain that economic growth can continue without limit, such speculation finds virtually no support in the historical record.

Envisaging Alternative Futures

Interest in alternatives to economic growth goes back a long way in the history of economics. In 1848, John Stewart Mill devoted a chapter in his *Principles of Political Economy*, an influential book for several decades, to

Figure 3–1. Material Footprint "Decoupling" in OECD Countries, 1991–2008

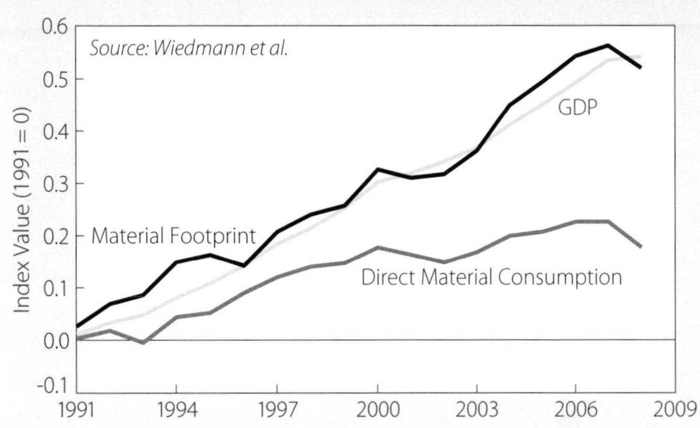

a consideration of the "stationary state." He was motivated to do so not because of a concern that economic growth could not continue, but because of what it was doing to life in Britain as he saw it. Although Mill's language comes from an earlier age, his sentiments are surprisingly modern: "I am not charmed with the ideal of life held out by those who think that the normal state of human beings is that of struggling to get on; that the trampling, crushing, elbowing, and treading on each other's heels, which form the existing type of social life, are the most desirable lot of human kind, or anything but the disagreeable symptoms of one of the phases of industrial progress. . . . The best state for human nature is that which, while no one is poor, no one desires to be richer, nor has any reason to fear being thrust back, by the efforts of others to push themselves forward."[20]

Mill was careful to acknowledge that, in developing countries, "increased production is still an important object," and that, "in those most advanced [countries], what is economically needed is a better distribution, of which one indispensable means is stricter restraint on population." Mill understood that in a stationary state, there would still be ample opportunity for technology to improve the quality of life, for example, by reducing time spent at work. He decried the extent to which humans were transforming land from its natural state and eliminating animals and plants that were not domesticated. He concluded his remarkable chapter on the stationary state by expressing the "hope, for the sake of posterity, that they [the population] will be content to be stationary, long before necessity compels them to it."[21]

Numerous writers, including several notable economists such as Keynes, Daly, and E. F. Schumacher, have cast their minds forward in contemplation of a future very different from their own times. A theme of particular interest is understanding what might be possible in advanced economies in the absence of economic growth and reductions in throughput. Would these economies collapse without growth? Would mass unemployment result? Could the existing institutions—in particular, financial institutions—survive without growth, and if not, what sort of changes might be required? What would be the implications for economic growth of strict limits on throughput?[22]

These and other questions are the focus of ongoing research based on an integrated approach that includes: 1) the financial system, with a central bank and commercial banks where money is created, credit is advanced, and interest paid; 2) the real economy, where resources are allocated and goods and services are produced and distributed; and 3) the throughput flows that link the real economy to the biosphere. This research, using simulation models and comprehensive databases, has shown how a different approach to labor productivity could improve the prospects for higher levels of employment, even in the context of declining economic growth rates. It also has shown that declining growth rates need not necessarily lead to high levels of inequality. At the local level, research has investigated the implications for communities of slow or no economic growth, focusing on enterprise, employment, investment, and finance.[23]

An earlier attempt to develop a model (termed "LowGrow") for scoping out alternative futures for Canada illustrates the kinds of insights that simulation models of economies can generate. In LowGrow, as in the economy that it represents, economic growth is driven by: net investment, which adds to productive assets such as machinery, buildings, and infrastructure of all types; growth in the labor force; increases in productivity; growth in the net trade balance (i.e., exports minus imports); growth in government expenditures; and growth in population. Low-, no-, and de-growth scenarios can be examined by reducing the rates of increase in each of these factors singly or in combination. One promising scenario is shown in Figure 3–2.[24]

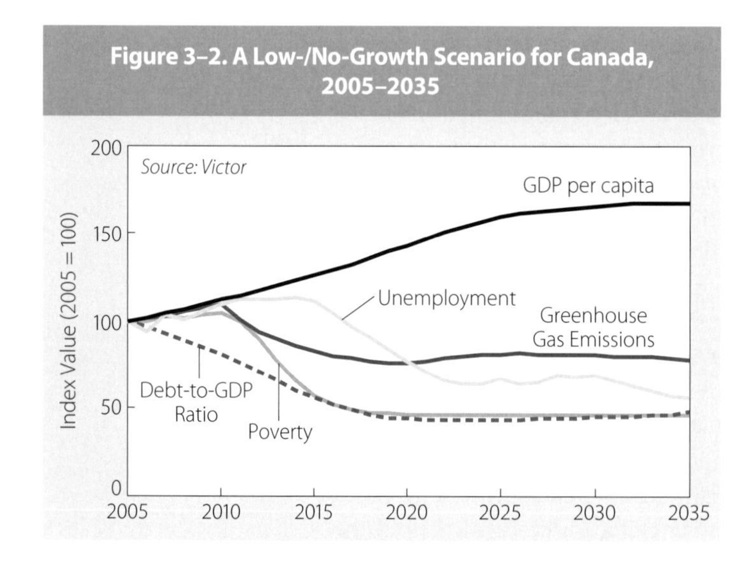

In this scenario, growth in GDP per capita slows until it levels off completely around 2030, at which time the rate of unemployment is 4.7 percent. The unemployment rate continues to decline to 4.0 percent by 2035, a rate that Canada has not seen for 50 years. By 2020, the United Nations poverty index declines from 10.7 to an internationally unprecedented level of 4.9, where it remains, and the debt-to-GDP ratio (a common measure of government fiscal performance) declines to about 30 percent, to be maintained at that level to 2035. Greenhouse gas emissions are 31 percent lower at the start of 2035 than 2005, and are 41 percent lower than their high point in 2010.

These results are obtained in LowGrow by slower growth in government expenditure, net investment and productivity, a small positive net trade balance, cessation of growth in population, a reduced workweek, a revenue-neutral carbon tax, and increased government expenditure on anti-poverty programs, adult literacy programs, and health care. There would still be plenty of opportunity in this scenario for technological advance, but it would be directed toward reduced throughput and away from activities that undermine sustainability.

The scenario in Figure 3–2 shows that even without economic growth, a range of desirable economic, social, and environmental objectives can be achieved. However, it would be a mistake to interpret this scenario as suggesting that zero economic growth, measured conventionally as increasing real GDP, should become an economic policy objective in its own right. From a sustainability perspective, what matters is an absolute reduction in throughput and land transformation degrading the soil and destroying habitat. These are necessary conditions for sustainability. Achieving them through strict controls that reduce throughput and rebuild ecosystems may well require a reduction in the rate of economic growth. It might also entail a period of degrowth until the economy's burden on the environment, in rich countries to start with, is sufficiently moderated. The main lesson from a scenario such as the one in Figure 3–2 is that we should not shy away from measures necessary for sustainability on the grounds that they will undermine economic growth.[25]

It is interesting to consider what life would be like under such a scenario. Much would depend on whether the scenario is adopted broadly, enthusiastically, and democratically, or whether the economy simply stagnates, as some fear is already happening, and nothing is done to alleviate the social stresses and friction that inevitably would result. In the sort of positive possibility envisaged here, there will be many changes. For example, rather than using gains in productivity to produce more goods and services, people would have more leisure time to spend with friends and family, and to participate in community life. Rampant financial speculation of the kind that brought widespread misery in 2008–09, and that still threatens, would be avoided through new banking structures and regulations. This would help facilitate a redirection of investment away from the endless and ultimately pointless search for social status from consumption, and toward increased investment in a wide range of public goods, such as community facilities, better infrastructure, and the protection and enhancement of air, water, soils, and ecosystems.

With much slower or no economic growth, it will no longer suffice to say that poverty will be eliminated through economic growth, a claim that has been proven wrong even on its own terms. In recent decades, most of

Alex

Coal-fired Navajo Generating Station near Page, Arizona.

the gains from economic growth have been restricted largely to the top few percent of the population. All Western countries have had experience with redistribution programs: progressive income taxation, inheritance taxes, income-support programs, universal health and education programs, low-income housing, and so on. More-equitable distribution can also involve greater use of cooperatives and more widely shared ownership. While redistribution through these kinds of instruments and institutional structures has fallen into disfavor in recent times, often in the name of economic growth, we expect them to play a vital role as we move toward a sustainable future.[26]

Conclusion

As the discussion above suggests, the pursuit of endless economic growth is a threat to sustainability. Most economists and governments are reluctant to come to grips with the implications of economic growth for the biosphere, preferring instead to hold out hope for absolute decoupling, to be delivered by a combination of technological change and a switch to a more service-based economy. Both of these avenues of change are important. But the existing evidence suggests that absolute decoupling of economic growth from throughput is unlikely: the historical record shows very little evidence of absolute decoupling, and assumptions about future decoupling are heroic at best.

History also shows that the pursuit of economic growth as a policy objective (and indeed as an object of academic study) is comparatively recent, going back only to the 1950s. Meanwhile, critiques of growth for growth's sake are longstanding, dating back to the 1800s. Achieving prosperity in its fullest sense is not at all synonymous with expanding the economy indefinitely. These considerations suggest that it is possible—and indeed desirable—to move public policy objectives away from the pursuit of economic growth, and toward specific goals that are more directly related to the well-being of humans and other species.

There are many good reasons for undertaking such a shift. One reason is that in much discussion about policy, economic growth is often used as a trump card. If protection of the environment threatens economic growth, then it is too bad for the environment. Hence the current interest in "green growth," with its false promise of even faster economic growth. Support for the arts, for sports, for child care, for less inequality, for better access to public goods, or for greater environmental protection all too often depends on whether a case can be made that it will promote economic growth, or trade, or competitiveness, or productivity, or some other growth-promoting consideration.

Our preoccupation with economic growth often has impeded action on issues that really will improve human well-being and the prospects for all life on Earth. This is the trouble with growth. If we insist on continuing to make economic growth the priority, we will deprive ourselves, and our descendants, of a sustainable future. It is time to remove the growth trump card. The pursuit of economic growth should no longer be a threat to sustainability.

CHAPTER 4

Avoiding Stranded Assets

Ben Caldecott

At the turn of this century, coal mining firms in Australia believed they had a bright future ahead. China's economy was headed into overdrive, and its leaders looked overseas for energy to fuel the unprecedented growth. Australian coal miners jumped at the opportunity. By 2013, Australia had become the largest supplier of coal to China, accounting for more than 30 percent of China's imports. Australian companies drew up plans to pursue 89 new mining projects that would more than double their country's coal output, largely for overseas markets like China's.[1]

Australian coal miners now have many reasons to fret over their expanded Asian market. As the Chinese government has stoked economic growth in recent years, it also has paid increasing attention to the country's environmental challenges, including the need to clean up China's notoriously polluted air. The government has passed a series of air pollution regulations, including an aggressive 2013 tax on the dirtiest elements of coal combustion, and in 2014, it agreed jointly with the United States to curb greenhouse gas emissions—moves that together are dampening Chinese demand for coal.[2]

Investors in Australian coal mining companies are already anxious. What will happen to their firms' expansion plans, and to the increase in company value that the planned investments represent? What will happen if Australian coal companies cannot find other customers to replace Chinese demand? Citizens and policy makers have questions, too. Would Australian society have been better off if another sector had been the target of investment capital? How can future investments be steered toward projects that support the country's interest in creating a sustainable economy?

In sum, what happens to investors, businesses, and society if Australian coal assets become "stranded assets"—assets that have suffered from unanticipated or premature write-downs, devaluations, or conversion to liabilities?[3]

The stranded assets dilemma is much larger than Australia, deeper than coal, broader than investors, and generated by factors beyond government

Ben Caldecott is a programme director at the University of Oxford's Smith School of Enterprise and the Environment, where he founded and directs the Stranded Assets Programme. He also is an adviser to The Prince of Wales' International Sustainability Unit.

policy. Across all continents, environmental and resource changes—from water scarcity to species losses to growing levels of greenhouse gases in the atmosphere—raise questions about the wisdom, from a societal as well as an individual investor perspective, of long-term investments that may lock economies into environmentally unsustainable economic activity. Visionary management of policies, companies, and investments is needed to ensure that new investments are consistent with environmental health and resilience, and that economies are weaned, smoothly and efficiently, off investments that are harmful to sustainability.

Risks That Lead to Stranding

Stranded assets can be caused by many different types of risks, but, increasingly, environment-related risks are stranding assets. This trend is accelerating, potentially representing a discontinuity that is able to profoundly alter asset values across a wide range of sectors. A stranded asset can start as a positive contributor to a firm's balance sheet, when environmental risks are underappreciated and unreflected in the asset's valuation. But as the risk becomes more apparent (sometimes over a short period), the asset becomes less attractive, to the extent that it may be abandoned before the end of its useful life. Today's financial and economic markets face extensive exposure to environment-related risks, many of which could result in stranded assets.

Stranded assets can include capital stock investments (such as extraction, production, and transport infrastructure), as well as current asset inventories (such as oil or mineral reserves, agricultural land, or natural resource inputs), that determine how firms may be valued. They are often large investments, characterized by fixed or sunk costs, and are relatively illiquid—they cannot quickly be converted to cash. Importantly, stranded assets can generate ripple effects well beyond their owners. (See Box 4–1.)

Table 4–1 shows a typology for different environment-related risks that could produce stranded assets. While the set of risks is diverse, they are

Box 4–1. The Tentacles of Stranded Assets

Stranded assets often produce collateral damage, in the form of lost physical, natural, social, and human assets. For example, fallowed cropland is a stranded economic asset if driven by, for example, overpumping of groundwater, a natural asset. Such fallowing, across a large number of farms, can weaken rural farming networks (a social asset) and lead to loss of income or employment for farmers (human assets). While most discussion of stranding focuses on stranded financial and economic assets, the environmental, social, and human assets that can be damaged by stranded financial and economic assets are of critical importance as well.

Table 4–1. Environment-related Risks That Could Produce Stranded Assets

Set of Risks	Subset	Examples of Assets Potentially Stranded
Environmental change	Climate change	Coastal zones more prone to storm surges and flooding
	Natural capital depletion and degradation	Forestry holdings
	Biodiversity loss and decreasing species richness	Pharmaceuticals
	Air, land, and water contamination	Farmland; tourist and recreational holdings
	Habitat loss	Holdings of species-sensitive real estate
	Freshwater availability	Cropland; certain industrial operations
Resource landscapes	Shale gas	Coal
	Phosphate	Farmland
	Rare earth metals	Electric motor manufacturers
Government regulations	Carbon pricing (via taxes and trading schemes)	Coal-fired power plants
	Subsidies (e.g., for fossil fuels and renewables)	Investments in fossil fuels and renewables
	Air pollution regulation	Power plants and other polluting infrastructure
	Disclosure requirements	Companies with poor sustainability performance
	International climate policy	Fossil fuel power plants
Technological change	Falling clean technology costs	Fossil fuel reserves
	Disruptive technologies	Distribution and transmission assets
	Electric vehicles	Car manufacturers
Social norms and consumer behavior	Fossil fuel divestment campaign	Fossil fuel companies
	Product labeling and certification schemes	Genetically modified agriculture
	Changing consumer preferences	Less energy-efficient products

Table 4–1. continued

Set of Risks	Subset	Examples of Assets Potentially Stranded
Litigation and statutory interpretations	Carbon liability	Fossil fuel companies
	Litigation	Owners and operators of polluting assets
	Changes in the way laws are applied or interpreted	Investments made under and dependent on previous legal interpretations

Source: See endnote 4.

not independent; correlations and connections among the risks are likely, although the extent of these interdependencies is yet to be determined and is an important area for future research. A critical issue for policy makers and financial institutions is to understand how this broad spectrum of risks might converge to imperil valuable assets.[4]

Fossil Fuels

The most publicized case of potential asset stranding is associated with upstream fossil fuel reserves: oil, natural gas, and coal left in the ground because of international climate change policy. Stranding of fossil fuels grew in public awareness after 2000 and gained widespread attention in 2011 with the publication of the Carbon Tracker report *Unburnable Carbon: Are the World's Financial Markets Carrying a Carbon Bubble?* and its popularization by U.S. environmentalist Bill McKibben. The report used simple but compelling logic to question the wisdom of continued investments in fossil fuels.[5]

The argument used by Carbon Tracker is as follows: to prevent global average temperatures from rising more than 2 degrees Celsius (°C) above preindustrial levels in the decades ahead—an internationally recognized goal—global carbon dioxide (CO_2) emissions need to be kept under 565 gigatons. Yet proven reserves of fossil fuels held by governments and private companies totaled five times this amount, or 2,795 gigatons. Thus, if the entire stock of proven fossil fuel reserves were burned, global temperatures likely would rise by more than is acceptable for climate stability. Stated differently, adhering to a carbon budget that limits temperature increase to 2°C requires that some 80 percent of proven fossil fuel reserves remain unburned—which would make them stranded assets.[6]

The idea that "unburnable" fossil fuel reserves could become stranded assets has been taken up by a number of high-profile actors and has helped spark a significant discussion on the risk of investing in fossil fuels. HSBC research concluded in 2012 that a global peak in coal consumption in 2020—a necessary condition for the transition to a low-carbon economy—would devalue existing share prices of coal assets on the London Stock Exchange by 44 percent. Ratings agencies like Standard & Poor's have begun to express concern that stranded asset risk could lead to credit downgrades. And the International Energy Agency acknowledges that a significant fraction of fossil fuel reserves is unburnable. Such concerns are not accepted unanimously, however, with fossil fuel companies claiming that they are based on oversimplified analysis. Each year, energy companies continue to spend more than $600 billion to find more fossil resources.[7]

Stranding of fossil fuels can happen to coal, natural gas, or oil, virtually anywhere in the world, and can be driven by any number of drivers—from climate concerns to regulations to the relative prices of other resources. Consider the following cases:

Coal in the United States. Between 2009 and 2013, 20.8 gigawatts (GW) of coal-fired power plants—some 6.2 percent of the 2009 U.S. coal fleet—were retired and another 30.7 GW were "slated" for retirement, with most estimates suggesting further retirements of 25–100 GW by 2020. The U.S. Energy Information Administration projects that 60 GW of coal will be retired by 2020, and a 2013 study by researchers with the Union of Concerned Scientists indicates that 59 GW of coal units are "ripe for retirement," in addition to the 28 GW already announced for retirement before 2025. Another 2013 study by Synapse Energy Economics takes into account a wider range of costs—including cooling water, water effluent controls, and coal ash—and estimates 228–295 GW of vulnerable capacity.[8]

The loss of coal capacity in the United States has several drivers:

- *Regulation*. In June 2014, the U.S. Environmental Protection Agency unveiled a new proposal to reduce CO_2 emissions from U.S. power plants 30 percent from their 2005 level by 2030. The biggest advances toward that goal can be made by reducing consumption of coal, an especially dirty fuel. The U.S. coal fleet produces 39 percent of the nation's electricity but is responsible for 74 percent of domestic power plant emissions.[9]
- *Low natural gas prices*. The shale gas boom has provided a cheaper and cleaner alternative to coal. A 2013 report from Bloomberg New Energy Finance predicts that U.S. natural gas prices will remain low (less than $5 per million Btu) until 2024, and forecasts that natural gas power plant capacity in the country will rise to 134 GW by 2030.[10]
- *Technological advance*. Renewable energy is an increasingly attractive alternative to fossil fuels. Wind energy costs have declined by some 80 percent in the last three decades, and the costs of solar photovoltaics

Max Phillips (Jeremy Buckingham MLC)

A shale gas well installation in Pennsylvania.

(PV) have fallen rapidly because of a steep drop in manufacturing costs. As a result, solar PV capacity in the United States has reached 8.9 GW, and rooftop PV installations are predicted to grow to 10 percent of the U.S. capacity mix by 2030.[11]

The U.S. government still expects 30 percent of electricity to come from coal in 2030. Already, however, significant investment has been stranded, and the coal industry has recognized that even more is at risk. Nick Atkins, chairman and CEO of American Electric Power Co., admitted in May 2014 that, "it's a critical issue for us not to strand all that investment that we made and secondly to make sure the grid can operate in a reliable fashion through this transition."[12]

Natural gas in Europe. Over the course of 2013, a large number of recently built, high-efficiency combined-cycle gas-turbine power plants across the European Union (EU) were closed prematurely or mothballed, including new, high-efficiency units—such as Statkraft's 430 megawatt (MW) Knapsack 2 plant, and Vattenfall's 1,300 MW Magnum unit—that were mothballed immediately upon commissioning. An estimated 51 GW of the EU's generation capacity is currently mothballed, and 60 percent of EU gas-fired capacity does not cover its fixed costs and could be closed within three years.[13]

The change of fortune for gas was driven by:

- *Decreased electricity demand.* As a result of the financial crisis, electricity demand dropped and has not recovered to pre-crisis levels.
- *Renewable energy deployment.* The intermittency of renewable energy and its priority in the order of power dispatch has affected capacity requirements and market price volatility.
- *Lack of a carbon price incentive.* The global financial crisis and subsequent slow down in economic growth resulted in less demand for carbon permits in the EU, which led to a significant and prolonged fall in European carbon prices. The lower the carbon price, the more competitive that coal is relative to natural gas.
- *Cheap coal from the United States.* As a result of the U.S. shale boom, profits of natural gas-fired power plants fell to the point of being uneconomic in comparison to coal power. Thus, while the fortunes of coal

have been declining in the United States, they were advancing in Europe.

Mothballing has resulted in significant write-downs on natural gas-fired power assets. The top 16 EU utilities reported €14.6 billion (around $17.5 billion) in impairments on generation assets over the course of 2010–12. Along with credit downgrades and the revision of dividends to preserve balance sheets, major utilities have curtailed planned capacity investments significantly, contributing to increasing fears about system security and the risk of blackouts in EU countries.[14]

Oil worldwide. Oil faces growing pressures on a number of fronts. For example, analysts estimate that if atmospheric concentrations of CO_2 are capped at 450 parts per million (the maximum level thought to be consistent with limiting temperature increase to 2°C), there would be $28 trillion less in oil revenues over the next two decades, compared with business as usual.[15] Other pressures include:

- *Climate regulation.* It has been estimated that global oil reserves can supply 1.8 times the oil allowable under a carbon budget determined by a 2°C goal for maximum temperature increase. Thus, if a global climate change agreement is reached, a sizable share of oil reserves will become stranded assets.[16]
- *Rapidly rising costs of access to oil.* As oil sourcing has shifted from low-hanging fruit to unconventional oil, such as shale oil and oil sands, and as companies turn to deep-water projects to produce conventional oil, extraction costs have increased markedly. Since 2000, oil industry capital investments have risen by 180 percent, largely to access these more challenging supplies, but the global oil supply has inched up only 14 percent. One analyst suggests that more than a third of potential production through 2050 will be high cost, requiring investment of $21 trillion and a minimum market sales price of $95 per barrel.[17]
- *Geopolitical risks.* Geopolitical risk is a growing concern with the increase in political instability. Through 2025, oil companies have $215 billion of capital expenditures planned in countries whose geopolitical risk Goldman Sachs rates as "high" or "very high."[18]

These factors increase the likelihood that newer oil holdings will not be exploitable and that their value will disappear from balance sheets.

Natural Capital

Economic assets also can be stranded as various elements of the biosphere—including land, air, water, and organisms—are polluted, depleted, or driven to extinction, diminishing their contributions to economic activity. (See Box 4–2.) But because the contributions of natural capital to economic health are often underappreciated and undercounted, the value of natural capital typically is poorly represented in decision making, and the connections between

Box 4–2. Nature's Contributions to Healthy Economies

Nature is an economic player, through its contributions of goods (such as wood, water, and air) and services (from crop pollination to flood control). Nature's goods and services, often called natural capital, provide ongoing life and resilience to ecosystems—as well as essential inputs to the world's economies.

But this asset base is being steadily eroded. The United Nations' 2005 *Millennium Ecosystem Assessment* calculated that 60 percent of the 24 ecosystem services that it assessed were being degraded or used unsustainably, suggesting widespread neglect of natural capital. This is costly: a 2013 report from TEEB found that in 2009, unpriced natural capital costs associated with primary production (agriculture, forestry, fisheries, mining, oil and gas exploration, utilities) and processing (cement, steel, pulp and paper, petrochemicals) totaled $7.3 trillion annually, the equivalent of 13 percent of global economic output that year.

Source: See endnote 19

natural capital and stranded assets often are hard to identify. Still, the linkages among natural capital, economic activity, stranded assets, and financial performance are becoming more apparent as losses of nature's goods and services grow in size and visibility.[19]

Environment-related risks and stranded assets can affect the financial system in several ways. A rapid devaluation of assets can spread across sectors as mispriced environment-related risks are reassessed. Generally, such contagion begins in a specific sector where mispricing is obvious and disproportionately large, then spreads to other sectors and jurisdictions. Nervousness about the future value of oil reserves could, for example, affect the value of companies that provide services to oil companies. And in China, water scarcity is prompting strong national controls on water use, which could close coal-fired power plants. Closures could ripple across global coal markets and affect major coal-exporting countries.[20]

A second transmission mechanism is the potential of natural capital degradation to trigger capital flight in resource-reliant economies, which in turn could lead to loss of income and economic instability. If a country or region experiences significant degradation of natural capital stocks and flows, capital may flow outward from the area as investors reallocate current and planned investments, divest from assets, or reorient operations to new nodes of production with a more reliable supply of natural resource inputs. In countries that are heavily reliant on industries built on natural capital stocks and flows, such capital flight could have serious macroeconomic consequences, affecting inflation, exchange rates, and international competitiveness, thus triggering fiscal and monetary policy responses.

A third pathway is through trade and global supply chains. The globalization of key commodity supply chains and the growing financialization of commodity markets have increased exposure to climatic shocks in particular. This process of "hazard globalization" represents a new dimension of environment-related risk transfer through which natural capital degradation could affect regional social, economic, and political volatility, which, in turn, may have implications for global financial stability.[21]

The environment's connection to the economy is illustrated using the case of the Arab Spring, the social uprisings in the Middle East and North Africa. An important driver was significant increases in the price of food, which, in turn, were linked to shifting weather patterns. Global wheat prices doubled from 2010 to 2011 in response to weather-induced supply shortages: drought and heat waves cut wheat production in Russia by 32.7 percent and in Ukraine by 19.3 percent, while cool, damp weather reduced output in Canada by 13.7 percent and excessive rain cut Australian output by 8.7 percent.[22] Tightened supplies prompted Russia to restrict wheat exports even as drought-stricken China turned to world markets to meet domestic demand. The spike in demand for wheat in international markets sharply affected major importers like Egypt, where a typical household spends 38 percent of its income on food, and this was a contributing factor to unrest in that country.[23]

Agriculture

Agriculture can undergo extensive asset stranding because of its dependence on natural capital. The 2013 Trucost/TEEB study *Natural Capital at Risk* pegs the total natural capital cost from agriculture—the "environmental and social costs of lost ecosystem services"—at $2.4 trillion per year. The report compared the costs of natural capital loss to revenues for various sectors, and agriculture was the sector most exposed. (See Table 4–2.) The natural capital cost of cattle ranching and farming, for example, is more than seven times greater than the revenues produced by these activities. The impact of agriculture on natural capital prompts the question: How would investment in agriculture be affected if the costs of natural capital were folded into the balance sheets of firms related to agriculture?[24]

Multiple environment-related risks could strand agricultural resources. A helpful way to think about risks to agriculture is to organize them by how quickly they can emerge and how long they might be a threat. (See Figure 4–1.) Economic drivers, such as regulations, are relatively fast-moving risks that can be put in place suddenly, perhaps through a change in government or the adoption of an international agreement. On the other hand, physical risks such as a changing climate tend to manifest themselves over a longer period. Risks also can be classified as short term or long term. Classic problems of the commons such as declining ecosystem services, water quality, and land degradation are long-term risks, while disease risks and changes in oil

Table 4–2. Direct Environmental Damage as a Share of Revenue for Select Economic Activities

Sector	Impact
	Natural Capital Cost as Percent of Revenue
Cattle ranching and farming	710
Wheat farming	400
Cement manufacturing	120
Coal power generation	110
Iron and steel mills	60
Iron ore mining	14
Plastics material and resin manufacturing	5
Snack food manufacturing	2
Apparel knitting mills	1

Source: See endnote 24.

Figure 4–1. Time Horizons for Environment-related Risks in Agriculture

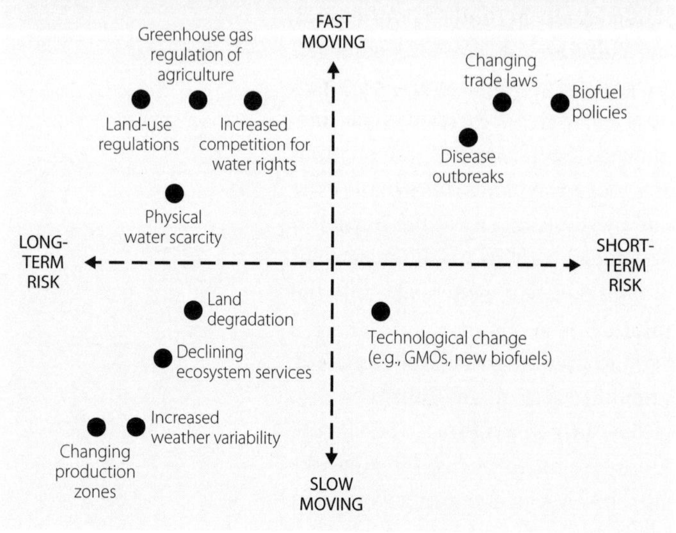

FAST MOVING
SLOW MOVING
LONG-TERM RISK
SHORT-TERM RISK
Greenhouse gas regulation of agriculture
Land-use regulations
Increased competition for water rights
Physical water scarcity
Changing trade laws
Biofuel policies
Disease outbreaks
Land degradation
Declining ecosystem services
Technological change (e.g., GMOs, new biofuels)
Increased weather variability
Changing production zones

prices are short-term in character.

Environment-related risks can have a significant effect on agricultural commodity prices. Three droughts in Australia between 2001 and 2007, and a heat wave in central Asia in the summer of 2010, drove down global stocks of several agricultural commodities (especially rice and wheat), which raised prices and led some major producing countries to institute export bans and export taxes. Meanwhile, government-mandated production of fuel crops, especially corn in the United States and edible oils in Europe, put further pressure on food supplies and food prices. Thus, international food prices underwent the longest sustained cyclical rise in real agricultural commodity prices of the past 50 years: by 2011, the FAO Food Price Index had reached more than double its 2000–02 level. The boom spurred an increase in the value of underlying agricultural assets such as farmland. The Savills index of average global farmland values, a leading global reference, has risen over 400 percent in the last 10 years.[25]

The boom in prices creates an attractive investment environment in the short term, but the environmental factors—especially climate change—are a cause for concern over the longer term. By 2050, it is very likely that climate change will increase the incidence of extreme drought, especially in the subtropics and low to mid-latitudes. Increased water stress is expected to affect twice the land area affected by decreased water stress.[26]

According to the Intergovernmental Panel on Climate Change, the share of the global land surface in drought is predicted to increase by a factor of 10 to 30, from around 1–3 percent of the land surface today to around 30 percent by the 2090s. The number of extreme drought events per 100 years and the mean drought duration are likely to increase by factors of 2 and 6, respectively, by the 2090s. Snow melt will come earlier and yield less in the melt period, leading to increased risk of droughts in snowmelt-fed basins in summer and autumn, when demand is highest. Water supplies from inland glaciers and snow cover are projected to decline in the twenty-first century, continuing a twentieth-century trend. This will reduce water availability during warm and dry periods—when irrigation is most needed—in regions supplied by melt water from major mountain ranges.[27]

The assets most vulnerable to increasing weather variability and changing production zones will be those characterized by high fixed or sunk costs and those of low liquidity that are tied closely to the value of land. Natural assets, such as farmland that is economically marginal in times of good weather conditions and high commodity prices, have been assessed as likely to be highly vulnerable to asset stranding from weather variability. This has particular relevance in the context of the current agricultural commodity and investment boom, which has stimulated new investment, some of which is likely to be unsustainable if commodity prices fall toward more long-term trends.

Other natural assets that may be highly vulnerable to asset stranding are poorly defined water entitlements attached to the land. If weather patterns change, resulting in reduced access to water, such informal allocations may be appropriated by higher-value users such as urban consumers.

Avoiding Stranding

The move away from polluting and resource-intensive economic activity has clear implications for existing assets and for future capital investment. While problematic for some firms and sectors, there is no reason why the stranding of polluting and inefficient assets should hinder economic growth and development.

Better understanding the process of value destruction and value creation in an economy can help policy makers secure an optimal rate of asset stranding given a country's level of economic development, targeted rate of economic growth, and sustainability concerns. Too little asset turnover could leave economies with insufficiently productive assets and significant environmental degradation, while too much could result in unmanageable losses for companies and financial institutions, as well as challenging social issues due to job losses and displaced industries. Employing the right tools in the right way is critical for transitioning away from at-risk assets. (See Box 4–3.)[28]

Another dimension related to securing an optimal rate of asset stranding is the avoidance of lock-in. Policy makers should avoid picking technologies and infrastructure that might quickly become outdated or inappropriate from a societal perspective.* An example could be new-build coal-fired power stations, given ever-increasing concerns about air pollution and water scarcity as well as the availability of cost-competitive alternatives. Lock-in of this kind is expensive for society as a whole and ties up capital that could be deployed productively elsewhere.

The profile of a transition pathway is also important. The value lost through asset stranding should ideally be more than offset by new value creation in other areas, and this should happen smoothly over time. This is preferable to a transition that is staggered or "lumpy," or one where value destruction overwhelms value creation, even if only temporarily. Without a smooth and gradual profile, it will be harder to secure political and societal support. An analysis of stranded assets can help to reveal the potential profile of a transition pathway, and also help to identify winners and losers across sectors. Identifying the groups affected (particularly negatively) can allow for the provision of targeted transitional help—another way of ensuring sustained support during a transition that might involve painful losses for some firms.

In terms of the financial system, better understanding the materiality of environment-related risks and the levels of exposure in different parts of the system will help regulators manage scenarios that could result in financial instability. Within financial institutions, revealing and better pricing environment-related risks will improve risk management and hedging, potentially improving system resilience as well as portfolio performance. Higher risk premiums for assets that are more exposed to environment-related risks also may have the added benefit of shifting capital allocations

* The corollary of this is that, in some cases, it might be better to "sweat" existing assets until viable long-term replacements can be found. In other words, instead of investing in an intermediate option that may need to be replaced relatively quickly, it could be better to defer investment.

Box 4–3. Tools for Retiring Assets

One way to shift away quickly from environmentally unsustainable assets is to pay owners to shut them down, using a tool known as a reverse auction. Bids represent the price that owners are willing to accept to give up an asset such as a logging permit, an oil well, or a coal-fired power plant. The lowest bid wins. Reverse auctions already have been used successfully to shrink fishing fleets in overfished areas, and to buy back pumping licenses in areas suffering from water stress. The funds used to pay for reverse auctions could come from special levies—for example, on electricity bills—or from foreign assistance (in the case of developing countries) or other sources.

If analysis of coal-fired power plants is any indication, the cost of such an approach may be manageable. In forthcoming research, the Stranded Assets Programme at the University of Oxford's Smith School of Enterprise and the Environment conservatively estimates the compensation bill for prematurely closing all existing sub-critical coal-fired power plants—the least-efficient plants—by 2025 at $47 billion in the United States and $106 billion in India. The cost tends to be lower in the United States because coal-fired power stations are older and owners are willing to accept less compensation for early retirement. As a result, even a small pool of funds to finance auctions could quickly close a large number of coal plants.

Premature closures would need to be accompanied by support for individuals and communities that are negatively affected, for example by job losses. Lost generation capacity also would need to be replaced by cleaner alternatives, and the rate of replacement likely would be the biggest constraint on the pace of any closure plan. These challenges need to be integrated into any broader coal-closure strategy.

Source: See endnote 29.

away from sectors that could be considered environmentally unsustainable, and toward assets that are more in line with a cleaner and more sustainable economy.

In addition to the implications for financial markets, environment-related risks and stranded assets will affect company strategy. Companies that are exposed to environment-related risk factors or that are dependent on clients exposed to these risks may need to adapt their business models. Exporters, particularly those exposed to environmental regulation in key export markets, could be particularly vulnerable.

Also at risk may be companies that depend on imported resources that could be affected by greater price volatility in international commodity markets due to environmental change. Ultimately, firms that are better able to manage emerging environment-related risks could secure significant competitive advantages over time. In a recent meta-analysis, 80 percent of the studies reviewed indicate that the stock-price performance of companies is influenced positively by good sustainability practices.[29]

CHAPTER 5

Mounting Losses of Agricultural Resources

Gary Gardner

Farmers in the U.S. state of California, the country's leading food-producing state, were troubled in 2014 as the state's worst drought in 109 years began to bite. Three years of poor rains had reduced supplies of surface water for agriculture by 36 percent, leading farmers to step up pumping of groundwater. But the additional pumping could not cover the entire shortfall of surface water, and some 173,000 hectares of irrigated land, nearly 5 percent of the state's irrigated farm acreage, had to be fallowed. The economic toll is estimated at $2.2 billion, including 17,000 lost jobs.[1]

Californians bounce back from the occasional drought, typically relying on wet years to replenish aquifers and snowpack, and to fill reservoirs. But under the new normal of climate change, droughts will likely be frequent this century, putting continuous pressure on the state's water resources. Aquifers will be harder to replenish, while snowpack is forecast to decline 12–40 percent by mid-century and as much as 90 percent by 2100, as warmer temperatures settle in.[2]

In addition to the water challenge, California continues to lose substantial swaths of farmland each year to urban development. Losses totaled more than 9,900 hectares between 2008 and 2010, the equivalent of more than 80 percent of the area of San Francisco. The double impact of water and land loss, paired with the loss of a stable and sufficiently wet climate, could reduce California's agricultural output at a time when demand for farm products, in the United States and globally, is on the rise.[3]

The loss of key agricultural resources such as water and land is hardly unique to California. Growing water scarcity is an increasingly urgent problem in regions as diverse as China, India, North Africa, and the Middle East. Farmland is lost or degraded on every continent, while "land grabbing"—the purchase or lease of agricultural land by foreign interests—has emerged as a threat to food security in several countries. Meanwhile, rising

Gary Gardner is a senior fellow and director of publications at the Worldwatch Institute.

concentrations of greenhouse gases degrade the quality of our atmosphere—a third resource pillar of bountiful agriculture.

These resource losses occur even as the Food and Agriculture Organization (FAO) of the United Nations (UN) projects that global agricultural demand in 2050 will be 60 percent higher than the three-year average for 2005–07. It is little wonder, then, that of 26 critical emerging issues identified by the UN Environment Programme in 2011, the challenge of ensuring food security was ranked third by scientists and second by major nonprofit groups and governments.[4]

Some countries turn to food imports to reduce their need for agricultural resources, but this solution can increase a country's vulnerability to disruptions in supply, a risk that may be unappreciated by policy makers. Fortunately, large reserves of food—crops that are wasted or that are used to produce other commodities such as biofuels or meat—are available to meet any shortfall created by resource loss. But the best and first solution is to preserve the resources that make global food production possible.

Keeping a Full Pantry Full

Global agricultural production has grown 2.5–3 times over the past half century and can rightly be described as cornucopian, with enough food produced to feed the entire human family, if it were distributed evenly. But complacency regarding the level of production is unwarranted for several reasons:

Persistent hunger. A large share of the human family—some 805 million people, or one out of every nine individuals—is chronically hungry. The challenge of ensuring that no one goes hungry becomes greater with population growth: the human family is projected to expand 36 percent by 2050.[5]

Grain-intensive diets. A poor person who sees an increase in income will typically add variety to his or her diet by supplementing grains and vegetables with sources of protein, typically from animals or fish, in the form of milk, cheese, meat, and eggs. The result can be a more diverse and interesting diet, but also an increase in the amount of grain required, as many livestock are fed grain.

Competition from biofuels. Production of biofuels (ethanol, biodiesel, and other fuels made from grains, sugar, and oilseeds) eats up nearly 40 percent of coarse-grain production in the United States, 50 percent of Brazil's sugar crop, and 80 percent of oilseed production in the European Union. Demand for biofuels has driven food price increases in the last decade as well: the FAO says that biofuels represent a "new market fundamental" that affects prices for all cereals.[6]

As demand for agricultural products grew by 2.2 percent per year between 1961 and 2007, the extent of arable land grew much more slowly—just

14 percent for the entire period. To meet demand, farmers intensified production, using mechanization, chemical fertilizer (in place of manure), new seed varieties, irrigation, and other advances to coax more from each hectare of land. Meanwhile, as fish populations collapsed in many ocean areas, fishers also turned to intensification, using aquaculture—or fish farming—to meet rising market demand. Farmers and fishers in the decades ahead will be challenged to continue to make each hectare and each fish farm yield ever-greater quantities of food. Yet rates of growth of agricultural production globally are only half the 3 percent annual rate seen in developing countries in the past.[7]

A Worrisome Waterscape

Agriculture commands upward of two-thirds of water withdrawals in most economies, and water can make land highly productive: irrigated farmland accounts for only 16 percent of arable land in use today, but it produces 44 percent of the world's food. Thus, expanding irrigated area is a proven high-leverage strategy for boosting food output. But water is increasingly scarce in many countries, and the potential to increase irrigated area is shrinking. For example, the FAO views water as the binding constraint for food production in all countries in the Near East and North Africa region, and says that water "remains a core issue that can no longer be tackled through a narrow sectoral approach."[8]

Sam Beebe

Center-pivot irrigation north of Umatilla, Oregon.

Indications of water scarcity around the world are manifold, and most have sobering implications for agriculture. A growing number of river basins are now considered "closed" (meaning that water for domestic, agricultural, and industrial uses competes with ecological needs), including the Indus, Yellow, and Amu and Syr Darya in Asia; the Nile in Africa; the Colorado in North America; the Lerma-Chapala in South America; and the Murray Darling in Australia. The potential for expanded irrigation in these basins is limited.[9]

But scarcity extends well beyond these major systems. A 2012 study analyzing scarcity in 405 river basins that contain 75 percent of the world's irrigated area documented severe water scarcity for at least one month per year in 201 of them (and somewhat lesser scarcity in other months). In 35 river basins that collectively are home to 483 million people, severe water scarcity is the norm for at least half of the year. In some cases, the scarcity translates to international tension. Egypt, for example, is pressuring Ethiopia to stop construction of a large dam on the Nile, the source of much of Egypt's fresh water. Egypt has vowed to "defend each drop of Nile water with our blood."[10]

Meanwhile, aquifers, which water some 38 percent of global agricultural fields, are increasingly overtapped. A 2012 study in the journal *Nature* estimated that some 20 percent of the world's aquifers are pumped faster than they are recharged by rainfall, often in key food-producing areas such as the Central Valley and High Plains of the United States, the North China Plain, the Nile Delta of Egypt, and the Upper Ganges of India and Pakistan. On the North China Plain, which produces about half of China's wheat, wells are now dug 120–200 meters deep, compared with only 20–30 meters a decade ago. And a 2002–09 study of satellite data revealed that the region encompassing Western Asia's Tigris and Euphrates river basins had lost 144 cubic kilometers of fresh water, nearly equivalent to the volume of the Dead Sea, and that 60 percent of the loss was caused by overpumping of aquifers. Similar depletions have been monitored in India, North China, North Africa, southern Europe, and the United States.[11]

At the economy-wide level, water availability can be measured in terms of renewable water resources per person. Table 5–1 shows the growing number of countries subject to various levels of water scarcity. It reveals that nearly half a billion people live under the tightest scarcity conditions ("absolute scarcity"), while more than 2 billion people—just shy of one-third of the global population—live in countries that experience some level of water supply challenge. These numbers could be conservative if climate change is factored in. A modeling effort published in 2013 found that climate change will raise the share of global population living under

Table 5–1. Number of Countries and Populations Subject to Water Supply Challenges, 1962 versus 2011

Water Status	Number of Countries		Population
	1962	2011	2011
Water stressed (< 1,700 m^3 per person)	8	22	1.9 billion
Water scarcity (< 1,000 m^3 per person)	9	15	389 million
Absolute scarcity (< 500 m^3 per person)	13	29	506 million
Total	**30**	**66**	**2.8 billion**

Source: See endnote 12.

conditions of absolute water scarcity by 40 percent compared with the effect of population growth alone.[12]

Absolute scarcity does not necessarily translate to poverty or suffering. Singapore, for example, is a prosperous country that is absolutely water scarce. But avoiding human deprivation under such conditions requires policies and practices that emphasize conservation—and leaves little room to absorb additional population growth or increases in water-intensive consumption. Indeed, as population expands in many water-tight countries, the number of people whose water availability is projected to fall below 500 cubic meters per person (the threshold for absolute water scarcity) will grow from just under half a billion in 2011 to some 1.8 billion by 2025.[13]

Not surprisingly, a high level of national water scarcity sometimes correlates with dependence on imported food. Although many water-challenged countries manage to feed themselves, the 23 most water-scarce nations for which grain import dependency can be calculated import an average of 58 percent of their grain needs, with 9 nations turning to imports for all of their grain. As water scarcity spreads, the number of countries turning to world markets for food could well increase.[14]

Already, many water-scarce countries are choosing to import food as a water management strategy, because the water burden can be shifted to exporting nations. The concept of "virtual water" is used to measure the water embodied in the production of goods, and gives a sense of net transfers of water across borders and oceans. Most of this transfer is in the form of agricultural goods: some 88 percent of global flows of virtual water are embodied in crops (76 percent) and livestock products (12 percent).[15]

The biggest net exporters of virtual water are the United States, Canada, Brazil, Argentina, India, Pakistan, Indonesia, Thailand, and Australia. The biggest net importers are North Africa and the Middle East, Mexico, Europe, Japan, and South Korea. Jordan, for example, imports virtual water (in the form of products and their processing) equivalent to five times its own yearly renewable water resources. In Malta, the external water dependency is 92 percent—meaning that 92 percent of the water used by residents of Malta (including water needed to produce imports to Malta) originates outside its borders. For Kuwait, this dependency is 90 percent; for Jordan, it is 86 percent; Israel, 82 percent; the United Arab Emirates, 76 percent; Yemen, 76 percent; Mauritius, 74 percent; Lebanon, 73 percent; and Cyprus, 71 percent. Some countries with high external water dependency, like the United Kingdom and the Netherlands, are not water scarce.[16]

Losses and Transfers of Land

Most of the 150–200 percent increase in agricultural output in the last half century was achieved through increases in yields, rather than by expanding the cultivated area (which grew only 12 percent over the period), because of the limited availability of land. Today, the FAO reports that essentially no additional suitable land remains in a belt around much of the middle of the planet, including countries in the Near East and North Africa, South Asia, and Central America and the Caribbean, many of which still have growing populations. Additional available land is found primarily in South America and Africa, but much of it is needed for ecological purposes or is of marginal quality.[17]

This makes preservation of the world's existing farmland crucial for global agricultural production. Yet farmland is increasingly under threat. Land is degraded or paved over on all continents, and rights to its use are being transferred across national boundaries. Combined with the loss of water for agriculture, and in the face of growing global demand, the potential toll on global harvests is consequential.

Unless undertaken with care, farming can lead to erosion, salinization, and other forms of degradation that reduce farmland's productivity. Two studies between 1990 and 2008 assessing degradation at the global level have suggested that some 15–24 percent of the world's land is degraded. In 2011, the FAO reported that 25 percent of land is highly degraded and another 8 percent is moderately degraded. One of the early studies measured degradation using a proxy yardstick, the decline in vegetative mass, which has serious climate implications. Less vegetative mass means that less atmospheric carbon is absorbed, leaving more carbon in the atmosphere to warm the planet. Thus, degraded land not only diminishes the productive capacity of farmland, but also weakens a key defense against climate change, which, in turn, further depresses food production (see below).[18]

Meanwhile, farmland is being scooped up in dozens of countries by foreign investment firms, biofuel producers, large-scale farming operations, and governments. Since 2000, agreements have been concluded for foreign entities to purchase or lease more than 36 million hectares, an area about the size of Japan. About half of this area is intended for use in agriculture, while 25 percent is intended for a mix of uses, some of which is agriculture. (Most of the remaining area is to be used for forestry.) Another nearly 15 million hectares are under negotiation. The bulk of the grabbed land is located in Africa, with Asia being the next most common region for acquisitions. (See Table 5–2.)[19]

The largest grabbers of land are often from countries that need additional production capacity, or whose corporations see profitable opportunities in land. But the largest source of land grabbing is the United States, a nation already rich in agricultural land. (See Table 5–3.) Target countries, on the other hand, are land-rich or water-rich, and some land is acquired as much for its access to water as for the land itself. Indonesia and the Congo, for

Table 5–2. Land Grabbed by Foreign Entities, by Region

Region (number of countries with grabbed land)	Grabbed Land Area	Share of Global Grabbed Land
	million hectares	percent
Africa (35)	20.2	55.6
Asia (15)	6.3	17.2
Oceania (1)	3.8	10.4
Latin America (16)	3.5	9.7
Europe (6)	2.6	7.1
Total (73)	**36.4**	**100.0**

Source: See endnote 19.

Table 5–3. Leading Investor and Target Countries for Land Investments

Investor Countries		Target Countries	
Country	Area Acquired	Country	Area Acquired
	million hectares		million hectares
United States	6.9	Papua New Guinea	3.8
Malaysia	3.6	Indonesia	3.6
Singapore	2.9	South Sudan	3.5
United Arab Emirates	2.8	Democratic Republic of the Congo	2.8
United Kingdom	2.3	Mozambique	2.2
India	2.1	Congo	2.1
Netherlands	1.7	Brazil	1.8
Saudi Arabia	1.6	Ukraine	1.6
Brazil	1.4	Liberia	1.3
Hong Kong (China)	1.4	Sierra Leone	1.3

Source: See endnote 20.

example, are water-rich countries that are among the most targeted countries for foreign acquisitions. In addition, contracts often do not take into account the interests of smallholders, who may have been working the acquired land over a long period.[20]

Land grabbing surged from 2005 to 2009 in response to a food price crisis, according to a 2012 report from the Land Matrix. Demand for biofuels is another driver. The 2007 Energy and Independence Security Act in the United States called for a fourfold increase in biofuel production by 2022, and a 2009 European Union directive had a similar stimulative impact. In addition, droughts in the United States, Argentina, and Australia drove interest in land overseas.[21]

An Abused Atmosphere

Our planet's atmosphere is another abused resource whose degradation could affect agricultural output. Changes in temperature, precipitation, and levels of atmospheric carbon dioxide (CO_2) and ozone all affect crop performance, in combinations that will increase output in some regions and decrease it in others. But the Intergovernmental Panel on Climate Change (IPCC) projects that the net effect on agricultural production will be negative. Higher temperatures are projected to lower crop yields and to increase the prevalence of weeds and pests. New, unpredictable rainfall patterns overall increase the risk of crop failures and, over the long run, production declines. The new patterns are expected to have much greater negative impacts in low-latitude (often developing) countries than in high-latitude (often wealthy) ones, and these divergent results are projected to widen over time.[22]

The IPCC noted in its *Fifth Assessment Report* in 2014 that crop yields could decline by 0.2–2.0 percent per decade over the remainder of the century, even as demand increases by 14 percent per decade. Across the various projections reviewed by the IPCC for major grains, outputs in the 2030–49 period range widely, from increased yields of 10 percent or more in the best 10 percent of projections, to losses of more than 25 percent in the worst 10 percent of projections, compared to the late twentieth century. And without action to stabilize climate, the probability of seeing effects that degrade yields increases steadily after 2050. From the 2080s onward, the IPCC describes the probability of negative yield impacts in the tropics as very likely across most emissions scenarios.[23]

Other studies suggest that the impact of climate change on agricultural output may be underestimated. A 2013 study in the *Proceedings of the National Academy of Sciences* layered a set of climate studies on top of hydrological studies to develop a fuller understanding of the impact of climate change on agricultural impact. The climate change studies, taken together, had suggested that the warming and precipitation could result in a loss of 400–2,600 petacalories of food supplies, or some 8–43 percent of present-day calorie levels. But once losses of irrigated area due to new rainfall patterns are added to the analysis, caloric loss increased by an additional 600–2,900 petacalories, essentially doubling the output losses expected from climate change by the end of the century.[24]

The impact of climate change on farming output could lead to higher food prices, although the effect of CO_2 fertilization influences the outcome. Without considering CO_2, global food prices are estimated to increase by 3 to 84 percent by 2050. Factoring in CO_2 fertilization (but not the dampening impacts of ozone, pests, and disease) produces projections ranging from a 30 percent decline to a 45 percent increase in prices by 2050. Already, the IPCC reports that some price spikes since its *Fourth Assessment Report* in 2007 are due to climate extremes in major producing countries.[25]

Food Imports: Too Clever by Half?

In the face of increased resource scarcity, more countries are turning to imports to meet their food needs. Such a strategy can help secure the calories needed by a country's population while conserving water: every imported ton of grain or meat saves thousands of liters of water domestically. For countries with dwindling water or land availability, food imports are a tempting way to escape rising resource pressures. But dependence on overseas suppliers for a basic requisite of life is risky, too.

The number of countries dependent on grain imports (defined here as importing 25 percent or more of domestic consumption) grew 57 percent between 1961 and 2013, to 77 nations—more than a third of the world's countries. (See Table 5–4.) Of these import-dependent countries, 51 (about a quarter of the world's countries) imported more than half of their grain in 2013, and 13 imported all of it. And in contrast to 1961, when no nation imported more than 100 percent of domestic consumption, by 2013 eight countries imported grain in amounts ranging from 106 to 127 percent of domestic demand, suggesting a perceived need to stockpile supplies. The number of grain exporting nations also grew over the period, but at a slower rate.[26]

Table 5–4. Number of Grain Importing and Exporting Countries, 1961 versus 2013

	1961	2013	Increase
	number of countries		percent
Grain Importing Countries			
100 percent dependent	11	13	18
More than 50 percent dependent	31	51	65
More than 25 percent dependent	49	77	57
Grain Exporting Countries	21	27	29

Source: See endnote 26.

Among developing countries, dependence on grain imports is greater than 50 percent in Central America, where land is relatively scarce, and in the Middle East and North Africa, where water is the chief constraint. (See Figure 5–1.) Sub-Saharan Africa imports about 20 percent of its grain, and the low- and middle-income nations of Asia import about 7 percent. Japan, with the wealth to outbid other nations in international markets, imports about 70 percent of its grain.[27]

A food import strategy is a logical response to resource pressures; outsourcing food production frees up land and water in huge quantities. But the strategy has two clear pitfalls. First, not all countries can be net food importers; at some point the number of countries demanding imported food could exceed the number supplying it. Already, many major supplier regions are themselves experiencing resource constraints, as the case of California demonstrates. Second, excessive dependence on imports leaves a country vulnerable to supply interruptions, whether for natural reasons (for example, drought or pest infestation in the supplying country) or political manipulation. An import strategy may now be unavoidable for some nations, but it should be considered only reluctantly by countries that can meet their food needs in more conventional ways. A better strategy may be to be vigilant in conserving agricultural resources wherever possible.[28]

Figure 5–1. Grain Import Dependence in Two Regions, 1960–2014

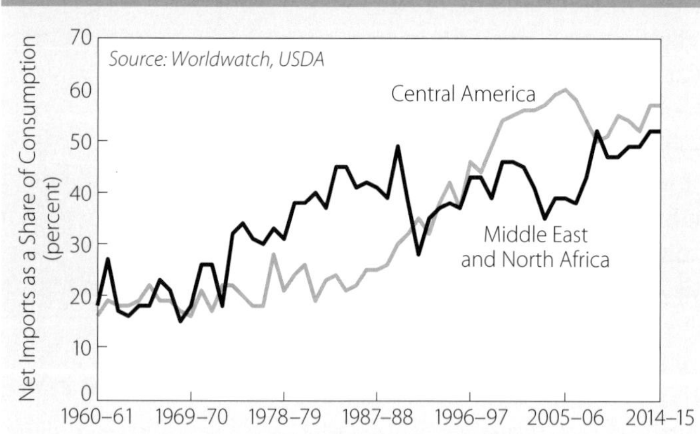

Prioritizing Conservation

As demand for agricultural goods increases, and as our planet's water and fertile land become more scarce and its atmosphere less stable, greater effort will be needed to conserve resources and to exploit opportunities for greater efficiency throughout the agricultural system. Fortunately, huge efficiency gains are available to farmers, food processors, businesses, and consumers. Taking advantage of these opportunities can help to ensure that food is available to the entire human family this century.

Combating food waste. A huge inefficiency in the global agricultural system—and therefore, a huge opportunity to conserve resources—is the 1.3 billion tons of food that the FAO says is wasted globally each year, a whopping one-third of annual global production. The FAO estimates that each year, consumers in Europe and North America waste 95–115 kilograms per person, while in sub-Saharan Africa and South and Southeast Asia they waste only 6–11 kilograms per person. Indeed, consumers in high-income countries waste almost as much food (222 million tons) as is produced in sub-Saharan Africa (230 million tons). In wealthier countries, food is more likely to be wasted in the home than in the supply chain leading to it, whereas in poor countries losses occur disproportionately at harvesting and during processing.[29]

Food waste can be avoided at many levels. At the farm and processing level, storage technologies can help to preserve harvested food, and they give farmers the flexibility of bringing their produce to market when prices are optimal. Comprehensive market infrastructure—wholesale, supermarket, and retail facilities—also help to ensure that food is forwarded efficiently to consumers who need it, although often at prices that are not fair to farmers. At the business level, "just-in-time" distribution systems can help ensure that restaurants and other businesses get their food only as it is needed. Small practices can make a difference as well: in cafeterias, tests have shown that not providing trays reduces waste by 25–30 percent while also trimming water and energy use. At the consumer level, education regarding food waste in wealthy countries can help change a culture of food waste to a culture of food stewardship, health, and nutrition.[30]

Importantly, as reductions of food waste lower demand for food, other resource savings are likely to result as well. Use of fertilizers, pesticides, water, and fuel would all decline, as would the volume of food rotting in landfills, which, in turn, would reduce generation of methane, a powerful greenhouse gas. In the United States, organic waste is the largest source of methane emissions.[31]

Increasing water productivity. Governments would also do well to increase water efficiency—through, for example, making drip irrigation available to farmers who can use it—and to set water productivity standards for farmers. Water footprint benchmarks developed in recent years for crops are a useful starting point for measuring the efficiency with which farmers use water. The benchmarks also are useful for the food processing industry, the biofuels industry, and, in the case of cotton, the apparel industry. The benchmark values can be used to measure performance and to monitor progress in achieving these targets. In addition, it may make sense to use water availability to guide selection of crops for water-scarce regions, with water-thirsty crops limited to water-abundant regions.[32]

The potential savings of farmers following global best practices are huge. Table 5–5 shows that if the listed crops all met the fiftieth percentile or better of modern water use efficiency, a quarter of the water used on these crops today could be saved. If crops were all grown at the top 10 percent of water efficiency, water savings for global production of these crops would reach 52 percent. In other words, achieving the water productivity of the best and near-best farmers would save half of the water used on these crops—a staggering achievement. The authors of the study suggest that because most high-level performance of the high-achieving farmers is the result of good management practices, rather than favorable climate or other natural factors, their success could be replicated widely across much of the world. Of course, part of good management is adequate access to the technologies, such as drip irrigation, and financing, that make high-yield agriculture possible.[33]

Table 5–5. Potential Water Savings from Increases in Water Efficiency in Agriculture

Crop	Global Total Water Footprint	Global Water Savings at Top 10th Percentile of Water Efficiency	Global Water Savings at Top 50th Percentile of Water Efficiency
	billion cubic meters per year	percent	
Wheat	964	64	25
Rice	881	60	18
Maize	648	51	35
Soybean	363	26	15
Sugar cane	254	43	21
Cotton	207	54	30
Barley	184	66	36
Sorghum	177	67	50
Millet	126	49	25
Potatoes	70	59	17
Others	2,750	47	23
Total	**6,624**	**52**	**25**

Source: See endnote 33.

Conserving agricultural land. Casual disregard for preserving the extent and quality of agricultural land no longer can be tolerated in most countries. A host of policy tools can be used to ensure that farmland remains farmland, from conservation easements to purchases of development rights. But stronger government action may be needed as well, including development and strict enforcement of agricultural zoning. In addition, governments need to be vigilant in preventing degradation of land by promoting conservation farming practices and discouraging careless use of marginal lands.

Reducing production of meat and biofuels. Two other large reservoirs of food that could be employed more efficiently for human consumption are the grains used to produce meat and the crops used to produce biofuels. Some 36 percent—more than one-third—of the world's grain harvest was used to produce meat in 2014. Fed directly to humans, this would feed many more people than it does in the form of beef, pork, chicken, or fish. Meat production is also water intensive, requiring thousands of liters to produce a kilogram of meat. (See Table 5–6.) A 2008 study found that the annual water requirement for food per person in China increased from 255 cubic meters in 1961 to 860 cubic meters in 2003, largely because of increased consumption of animal products.[34]

Table 5–6. Water Needed to Produce Various Types of Meat

Meat	Water Requirements for Production		
	liters per kilogram	liters per calorie	liters per gram of protein
Chicken meat	4,325	3.00	34
Pig meat	5,988	2.15	57
Sheep/goat meat	8,763	4.25	63
Bovine meat	15,415	10.19	112

Source: See endnote 34.

Healthier diets that reduce meat consumption are a logical response to the resource intensity of meat. Research comparing shifts in diet toward guidelines set by the World Health Organization have found that water footprints could be reduced by 15 to 41 percent, with the higher values achieved in industrial countries. In these countries, a vegetarian diet is estimated to reduce water consumption by 36 percent.[35]

Meanwhile, the U.S. government projects that between 2013 and 2022, biodiesel production will grow by 30 percent, and ethanol by 40 percent, in the seven countries that dominate the biofuel sector. Biofuels have eaten up a share of the surplus production that long characterized global agriculture, and that kept food prices low for much of the past half century. Key to eliminating this distortion to the global food system is reversing government mandates for biofuels production, which are now present in some 60 countries.[36]

Ethicizing international food markets. As the number of people living in countries that import more than a quarter of their grain use surpasses perhaps 1 billion in the decades ahead, food trade will become an indispensable nutritional lifeline. As such, food trade cannot be treated as just another exchange of goods, and food cannot be treated as just another commodity. Full development of the concept of the right to food, and its embrace by all governments, will be needed to ensure that the flow of food is never interrupted. The FAO advanced this concept in 2004 with the adoption of the Right to Food Guidelines, and at least 28 nations have an explicit right to food in their national constitutions. Codifying a right to food in international trade agreements so that, for example, food cannot be withheld for political reasons, may be required.[37]

In sum, conserving the very base of food production—the land, water, and climate that make crop growth possible—is essential to ensure that the world's farmers continue to produce enough food for everyone. Where resources already are scarce, reservoirs of food can be tapped for broader distribution and utilization. And political assurances guaranteeing that agricultural plenty is not blocked from dinner tables worldwide can ensure that food attains a sacred status in a globalized world. In these ways, a world under growing resource pressure can continue to ensure that food is available for all.

CHAPTER 6

The Oceans: Resilience at Risk

Katie Auth

In Herman Melville's 1851 classic, *Moby-Dick*, the deranged and obsessive Captain Ahab travels the world in pursuit of a singular white whale. As Ahab's crew sails on—facing storms, sharks, and, at times, desolate isolation—Melville transforms the ocean into a character in and of itself. The source of both great wealth and great danger, the sea becomes a symbol of humanity's simultaneous industriousness and utter powerlessness.

No matter what scientific or economic achievements humanity may make, Melville wrote, the ocean would forever be capable of taking lives and destroying even our most impressive technological creations. In the twenty-first century, with climate change contributing to rising sea levels and larger, more powerful storms, this prophecy remains all too relevant, enhanced by the irony that the greatest threats posed by the ocean result in part from our own actions. We are reminded of the ocean's power during tragic events like the 2004 tsunami that devastated parts of Southeast Asia or 2012's Hurricane Sandy in the United States, and when considering the existential threat that sea-level rise poses to small-island states and even major coastal cities. However, above all things, the ocean sustains us—powering both economies and critical ecological cycles.[1]

By the mid-1800s, when Melville published his novel, the ocean had fueled tremendous economic growth throughout the young United States. Whalers like those of Ahab's crew sailed from Gloucester, New Bedford, and other New England towns, returning home to supply consumers and manufacturers with whale oil for lighting and other applications. Fishermen voyaging to and from famous grounds like the Grand Banks of Newfoundland supplied fish for food and trade, and supported early American economies through the growth of industries like shipbuilding and salt mining. A "sacred cod"—carved from pine—still hangs in the Massachusetts State House, highlighting the region's deep-seated relationship with the fishing industry and with cod in particular, a cultural symbol of prosperity and identity.[2]

Katie Auth is a former research associate at the Worldwatch Institute whose current work centers on climate-resilient development strategies and international cooperation.

Preparing the cod catch for salting, off Newfoundland, 1891.

Even as we drew on marine resources for food, fuel, and raw materials, however, the ocean remained a largely mysterious and dangerous entity. Our sense of the ocean's power and omnipotence—combined with scientific ignorance—contributed to an assumption that nothing we did could ever possibly impact it. We assumed that waste could be dumped into the sea without consequence, that we could hunt fish for food and whales for oil without making a dent in their numbers, and that we could plunder the sea's riches to no end. Over the years, scientists and environmental leaders have worked tirelessly to demonstrate and communicate the fallacy of such arrogance. Thanks to their efforts—and to research conducted by countless other agencies, universities, and explorers—we now understand much more about the oceans and the services they provide than Melville did when he called the sea a *terra incognita*.

Yet more than 160 years later, we still know startlingly little about the diverse ecosystems covering nearly two-thirds of the earth's surface. By some estimates, we have explored less than 5 percent of the global ocean. However, for all our ignorance, one thing has become clear: the ocean is not, in fact, invulnerable. The same dual quest for fish and fuel that drove crews like Ahab's to sea has imposed enormous and multiple stressors on coastal and marine ecosystems, hindering their capacity to maintain resilience. While major environmental stressors such as overfishing and climate change each result in distinct negative impacts, it is their complex intersection that poses the greatest threat to marine ecosystems.[3]

As our negative impact on the oceans has grown, so has our understanding of the myriad ways in which the health of the marine environment determines our own. We depend on the ocean to supply us with food and oxygen and to maintain a balanced carbon cycle. In recent decades, its capacity to supply these ecological services has come under immense stress as a result of human activities. Restoring the ocean's health and rebuilding its capacity to withstand environmental stressors means taking rapid and innovative steps to change the way we produce and consume seafood, address waste and runoff, and generate energy. It also means dedicating the resources necessary to continue critical research into marine science and climate impacts, expand ocean education, and increase public awareness of how everyday activities affect coastal and marine ecosystems—and why it matters.

Unsustainable Fishing

Worldwide, fish represent the main source of animal protein, essential micronutrients, and fatty acids for an estimated 3 billion people. Dietary reliance on seafood is particularly high in developing countries and so-called Low-Income Food-Deficit Countries (LIFDCs), marking the critical importance of fish for food security. Fisheries also represent a key economic sector in both developing and high-income countries, employing men and women in marine and inland fisheries, aquaculture, and processing—including some 800,000 people in Egypt alone. (See Table 6–1.) In 2010, the U.S. commercial fishing industry accounted for 1.5 million jobs and yielded more than $45 billion in income.[4]

Table 6–1. Employment in Fisheries and Aquaculture in Selected African Countries, 2011

Country	Total Employed	Share of Women Employed
		percent
Egypt	796,400	1
Democratic Republic of the Congo	376,275	51
Mozambique	374,027	1
Mali	354,060	8
Benin	214,202	38
Malawi	173,328	9
Madagascar	166,013	5
Senegal	129,090	30
Kenya	105,132	33
Tanzania	517,126	28

Source: See endnote 4.

According to the United Nations Food and Agriculture Organization (FAO), global food fish supply (including both wild capture and aquaculture) has increased at an average annual rate of 3.2 percent over the past 50 years, roughly double that of the human population. (See Figure 6–1.) This reflects a combination of factors, including population growth, rising incomes, and the rapid expansion of aquaculture, which has accelerated even as wild capture fisheries have slowed.[5]

Since peaking in the 1990s, production from the world's marine fisheries

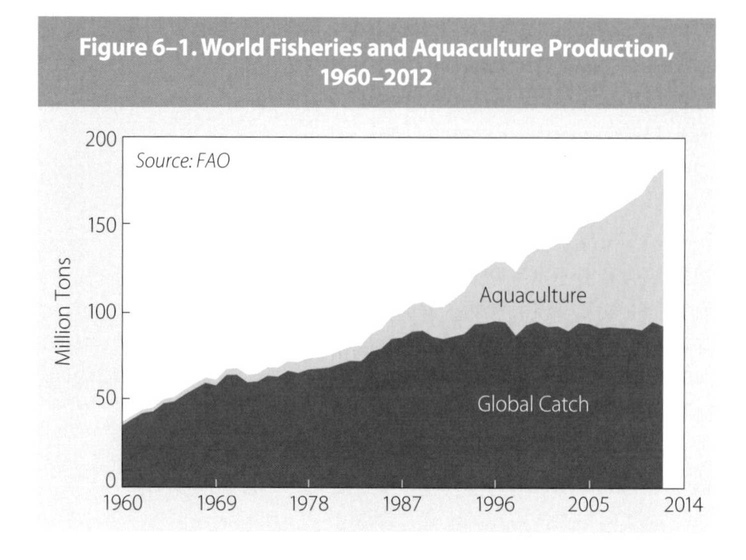

Figure 6–1. World Fisheries and Aquaculture Production, 1960–2012

has generally leveled off, according to FAO statistics. This likely reflects, at least in part, deteriorating ecological conditions, since catch levels have been buoyed by increasing effort and the expansion of fishing into new and deeper parts of the ocean. Although it remains difficult to accurately assess the size and health of marine species populations, estimates point to the severe impacts of overfishing. Between 1974 and 2011, the global share of assessed marine stocks considered by the FAO to be fished "within biologically sustainable levels" fell from 90 percent to 71 percent. Of that 71 percent, 86 percent are fished to capacity or "fully fished," meaning that there is no room to increase catch levels sustainably.[6]

Grim projections for the future of fisheries underline the cause for concern. A 2006 study examined the impacts of biodiversity loss on fisheries and projected the global collapse of exploited fish stocks by 2048 under a business-as-usual scenario. Ongoing controversy over such dire projections stems, in part, from the fact that available data on the numbers and types of fish caught at sea, as well as the health of specific stocks, remain vastly incomplete, reflecting the largely extrapolatory nature of fisheries science, inaccurate and misreported catch, and the fact that many fisheries—particularly in developing countries—are not formally assessed.[7]

Despite incomplete ecological data, it is clear that overfishing is a social, political, and economic challenge, rooted firmly in our collective failure to manage fisheries in a way that safeguards marine resources and ecological services. Particularly since the rise of industrial fishing, poor governance and failed attempts at fisheries management have contributed to fisheries collapse, while attempts at management reform remain a tense and socially charged issue in many parts of the world.

One of the United States's most iconic fisheries—the Gulf of Maine cod that inspired the Massachusetts State House's "sacred" pendant—went from being a symbol of wealth and seemingly inexhaustible bounty to one of mismanagement and collapse in only a few decades. In 2012, the U.S. Department of Commerce declared the fishery a federal disaster for the second time in 20 years. As of 2014, the Northeast Fisheries Science Center reports that Gulf of Maine cod stocks are at just 3 percent of what is needed to sustain a healthy population.[8]

According to an assessment of factors contributing to the fishery's struggles by the Pew Charitable Trusts, a history of mismanagement explains much of the crisis. First, the region repeatedly delayed setting hard catch limits on the total number of fish that could be brought ashore, despite evidence that the fishery faced major challenges. In 2010, when science-based limits were finally imposed, an inadequate system of monitoring and reporting negated much of the intended benefit, resulting in unreported bycatch (the unintentional capture of non-target species), discards, and illegal landings. In 2010, a study estimated that 12–14 percent of the total groundfish catch in New England was taken illegally. In addition to the ecological damage, this means that efforts by policy makers, scientists, and stakeholders to develop targeted and effective management plans do not reflect a full or accurate understanding of the situation. Finally, Pew noted, a failure to protect crucial habitats, especially those where fish spawn or juveniles seek shelter, has weakened the cod population's resiliency and its ability to recover, even with stricter catch limits.[9]

This kind of chronic management failure often reflects the complexity of governing a resource that is simultaneously difficult to track and monitor, economically important, and intertwined with regional and social identity. Iceland, a North Atlantic island nation that is heavily dependent on the ocean, often is lauded for its early decision to privatize its fisheries and implement a system of Individual Transferable Quotas—which operates much like a cap-and-trade policy for emissions—to rationalize the industry and protect the resource. Decades after the policy was implemented, however, it remains a source of fierce debate because of its effects on regional development, wealth disparity, and the perceived privatization of what many consider to be a public good and a source of personal, regional, and national identity.[10]

If implemented effectively, fisheries management methods—including aggressive catch quotas, community management techniques, targeted fishing gear that limits wholesale destruction of marine ecosystems, and economic incentives—hold the potential to reverse the overfishing trend and rebuild marine biodiversity. Although fisheries management has improved in many parts of high-income countries, the situation in Southeast Asia, Central America, Africa, and the Indian Ocean continues to worsen as both small-scale and industrial fishers operate largely without oversight or restriction.[11]

Because individual species play specific roles within an ecosystem, overfishing a particular species can have significant impacts on food availability and predator-prey relationships, altering the entire ecosystem. The decline of global shark populations presents a particularly stark example. Along

NOAA

An agent with the U.S. National Oceanic and Atmospheric Administration counting confiscated shark fins.

the eastern coast of the United States, the abundance of most local shark species declined by at least half during the 1990s–2000s, with thresher sharks diminishing by as much as 75 percent. Eleven species have declined to the point of "functional elimination," meaning that they can no longer fulfill the ecological role of top predator. A 2007 study documented the cascading effects of this collapse on local ecology, including a corresponding rise in the number of cownose rays and their increased consumption of bivalves such as clams, oysters, and scallops. In addition to posing an economic challenge to these fisheries, this may render bivalve populations more vulnerable to environmental stressors and less responsive to crucial restoration efforts.[12]

According to the FAO, annual global shark catches tripled between 1950 and 2000, when they reached an all-time high of 893,000 tons. Much of this has been driven by Chinese demand for shark fin soup, a highly valued gourmet item and a cultural symbol of luxury and status. Although the Chinese government announced in 2012 that it would no longer allow shark fin soup to be served at official banquets, the fishery's environmental legacy remains. As of 2011, global shark catches had fallen by 15 percent from their 2000 peak. While this reduction may reflect in part the impact of conservation measures designed to reduce mortality and avoid bycatch, the FAO notes that it also likely reflects "the overall declining abundance of fished sharks."[13]

The most common shark conservation measure, implemented at both national and regional scales, has been a ban on discarding shark carcasses after cutting and storing the valuable fins. In theory, this rule capitalizes on a vessel's storage limitations to reduce the maximum number of sharks that can be caught during a single trip. However, the difficulty of monitoring and enforcing the compliance of boats at sea limits the overall effectiveness of such bans. Despite concerted attempts to reverse overfishing, threatened shark species have not yet begun to recover.

As global catches of wild fish have leveled out over the past two decades, aquaculture has grown quickly in scale. Global production (excluding the harvest of aquatic plants) increased by a third between 2007 and 2012. China in particular has expanded its aquaculture production dramatically, with per capita fish consumption rising about 6 percent per year to 35.1 kilograms in 2010, compared to a global average per capita consumption of 15.4 kilograms.[14]

Aquaculture presents opportunities to address food security and to facilitate seafood consumption while reducing fishing pressure on wild stocks and perhaps consumption of other environmentally destructive sources of animal protein as well. But concerns about the industry's sustainability remain, including the wild fish that go into the production of fish meal and fish oil used to feed cultivated fish, as well as the impacts on wild populations, including the potential for transfer of disease and for escaped fish to alter wild gene pools.[15]

Climate Change

With the discovery and extraction of large quantities of "rock oil," beginning in Pennsylvania in 1859, petroleum began supplanting the use of whale oil in lamps in the United States. From this small beginning, the world's growing appetite for fossil fuels raised carbon emissions, and their impact on the oceans increased. The oceans are a major global carbon sink, sequestering in sediments and the deep ocean about a quarter of the carbon dioxide (CO_2) emitted each year as a result of fossil fuel consumption, land-use change, and cement manufacturing.[16]

As emissions increased, so did the oceans' rate of uptake. The oceans' natural sequestration of carbon may be buffering us temporarily from some of the worst potential impacts of human-induced climate change. A 2014 study suggests that ocean circulation has been depositing atmospheric heat in the deep Atlantic, possibly explaining why the rise in global surface temperature has slowed since 1999 despite continued greenhouse gas emissions—a phenomenon sometimes referred to as the "global warming hiatus" that has inspired climate change skeptics. However, evidence suggests that as the ocean becomes saturated with CO_2, its rate of uptake will slow, a process that perhaps has already begun.[17]

Unfortunately, carbon absorption is also profoundly changing the fundamental physical, chemical, and biological properties of the seas. In recent years, the complex and interconnected impacts of climate change on marine organisms and ecosystems have garnered increased attention from scientists, policy makers, and activists. However, these effects and the ways in which they interact are not yet fully understood. Two major climate impacts, temperature increase and ocean acidification, reveal examples of this interplay.

Warming. All organisms have a limited temperature range within which they can function and thrive. Over the past 40 years, the upper 75 meters of the world's oceans have warmed at an average annual rate of more than 0.1 degrees Celsius (°C). In 2012, sea-surface temperatures rose to their highest levels in 150 years, with 2013 ranking a close second. This places significant pressure on marine organisms, which must respond—if they can—by adapting (through active migration or passive displacement) or acclimatizing (shifting thermal tolerance).[18]

Rising ocean temperatures already are having immense and complex impacts on marine ecosystems. Temperate species—including, for example, pelagic fish species (those living in the open ocean) and the marine mammals that prey on them—have been documented shifting toward the poles in both hemispheres, although there is little agreement regarding the degree to which such shifts reflect only warming or, more likely, a combination of factors including warming, fishing pressure, and pollution. A 2009 study found that 24 of 36 assessed fish stocks on the U.S. northeast continental shelf were shifting north and/or moving into deeper water. As species shift their distribution poleward, seasonal migrants into the Arctic—including fin, minke, gray, killer, and humpback whales—may compete increasingly with species adapted to live with sea ice (such as narwhals, bowhead whales, and belugas).[19]

Ocean warming does not occur in isolation; it interacts with, and in some cases amplifies, the negative impacts of other human activities on marine ecosystems. In recent decades, the ubiquitous presence of microplastics in marine environments has raised widespread concern. Humans release small plastic fragments into the oceans in a variety of ways. Larger items of plastic debris, discharged from rivers, blown from land, or lost at sea from fishing boats or cargo ships, ultimately fragment and break down into small pieces. We also discharge microplastics from cosmetic products and clothing fibers into the wastewater system, where they ultimately enter the environment.[20]

NOAA

Marine debris on a beach in Kanapou Bay, Hawaii.

Once released into the ocean, plastic debris is transported widely by tides and currents, and ingested by mammals, fish, birds, and invertebrates. Upon ingestion, microplastics may release harmful chemicals and toxic additives such as plasticizers, flame retardants, and antimicrobial agents into biota. Ingestion of even small quantities of microplastics has been shown to interfere with physicological processes in marine worms, affecting their ability to store energy.[21]

Climate change may amplify this threat in unexpected ways. As sea ice forms, it concentrates natural particulates from the water column. A recent study found that concentrations of microplastics in Arctic sea ice, even in remote locations, can exceed by several orders of magnitude the concentrations found in notoriously polluted surface waters like the Pacific gyre. As sea ice melts in response to warming temperatures, these microplastics could be released into the sea, posing additional threats to marine ecosystems.[22]

Ocean acidification. As a natural carbon sink, the oceans have absorbed about a quarter of all anthropogenic (human-caused) CO_2 emissions released into the atmosphere to date, with significant impacts on ocean chemistry. When seawater absorbs CO_2, a series of chemical reactions reduces the water's pH (i.e., increases its acidity), lowers the concentration of carbonate ions, and reduces the saturation level of calcium carbonate minerals. Because calcium carbonate forms the basis for shells and skeletal structures in many marine organisms (for example, in oysters, clams, corals, and sea urchins), this poses a distinct threat to marine life and the food web. Elevated concentrations of CO_2 also have been found to interfere with neurological processes in fish, resulting in behavioral changes that may impede survival.[23]

Since the Industrial Revolution, the acidity of open-ocean surface waters has increased by about 30 percent. If emissions continue at current levels, ocean acidity in surface waters could increase by almost 150 percent by 2100, creating a marine environment unlike anything that has existed in the past 20 million years. Although ocean acidification can be buffered through natural processes, including erosion and the dissolution of calcium carbonate from sediments, these longer-term mechanisms likely are unsuited to cope with the rapidly rising acidity resulting from anthropogenic CO_2 emissions.[24]

The rate of acidification varies geographically. In the polar regions, acidification can be exacerbated by excess precipitation or ice melt—both projected to increase as a result of climate change—because these processes reduce salinity and decrease the concentration of substances needed to buffer the acidification process. In addition, high-latitude oceans naturally contain lower concentrations of calcium carbonate minerals and therefore are more vulnerable to ocean acidification because additional losses of calcium carbonate impose a greater relative change. Individual species' responses to ocean acidification vary as well, with growth stimulated in some animals and hampered or unaffected in others. Overall, an organism's ability to withstand changes in acidity depends on other factors that support overall resilience, including quality of food and fitness.[25]

The cumulative threat posed by climate change to marine organisms and ecosystems is not yet fully understood. Part of the challenge in understanding climate impacts is that distinct climate-related factors often reinforce each other. For instance, ocean warming reduces oxygen solubility and enhances organisms' oxygen demand, thereby exacerbating hypoxia, producing increased CO_2, and accelerating acidification.[26]

Links Between Unsustainable Fishing and Climate Change

The ocean has repeatedly proven to be remarkably resilient. However, the combined stresses of human activities like overfishing and climate change now pose distinct and intensified threats to marine systems. Over the years, overfishing has reduced the age and size structure of fish populations, reduced the prevalence of large and adult individuals, and limited populations' capacities to withstand other environmental fluctuations.[27]

As climate impacts such as rising temperatures and increased acidity worsen, marine ecosystems and individual organisms that already are weakened by overfishing become less resilient and more vulnerable to disruption, especially because environmental change is occurring so rapidly. As the Intergovernmental Panel on Climate Change (IPCC) acknowledges: "The limits to acclimatization or adaptation capacity are presently unknown. However, mass extinctions occurring during much slower rates of climate change in Earth's history suggest that evolutionary rates in some organisms may not be fast enough to cope."[28]

Ocean warming can have severe impacts in places where overfishing already has placed marine populations under stress. The Gulf of Maine is warming particularly quickly—faster than almost any other ocean waters. This rapid change, equivalent to about 0.55°C every two years over the past decade, is upending both marine ecosystems and the human communities that rely on them. The Pew report documenting the factors contributing to the collapse of cod in the region notes the interplay between overfishing and climate. The management failures cited in the report, including long-term overfishing and habitat destruction, have contributed to making the species increasingly vulnerable to climatic stressors and less responsive to restoration efforts. This may help to explain why the Gulf of Maine has struggled to build back cod stocks despite reductions in catch quotas.[29]

Often, this kind of complex interplay between various environmental stressors makes it difficult to understand and identify the triggers of large-scale ecological change. For example, the northward shift of species such as white hake and Atlantic herring, dietary staples of Atlantic puffins and terns, has forced these seabird populations to change their behavior. Dependent on proximity to nesting grounds, they cannot easily follow fish northward or out into deeper waters. Faced with a limited food supply, puffins in the Gulf of Maine have been documented trying to adapt by feeding their chicks replacement prey such as butterfish. The larger-than-usual fish are difficult for chicks to swallow, however, and puffin colonies face dwindling survival rates.[30]

Puffins in northern Iceland.

Similar ecological dynamics are playing out in Iceland, where the converging currents of the Atlantic and Arctic oceans and the island's rocky cliffs have long made the shoreline an ideal nesting ground for seabirds. However, on the Westman Islands, home to the world's largest Atlantic puffin colony, breeding has failed since 2005, mimicking trends in other puffin hotspots such as Scotland, the Faroe Islands, and Norway. Scientists cite a combination of warming ocean temperatures (which disrupt the puffins' major source of food) and pollution, especially mercury released by coal-burning power plants.[31]

Climate-induced shifts in marine ecosystems will pose a significant challenge for human populations. The IPCC projects that a 2°C increase in global temperature by 2050 would result in annual losses of $17–$41 billion from commercial fisheries. As the geographic range of species shifts, geopolitical arrangements such as international fishery agreements may be challenged. In the United States, Alaska's immense coastline produces half the country's total commercial catch, supporting 90,000 full-time-equivalent jobs in 2009. The state's most productive fisheries are located in areas projected to undergo rapid and significant changes in temperature and acidity.[32]

Globally, the human populations at particular risk from climate-related changes in marine ecosystems are those with fewer resources and lower adaptive capacity, such as communities on the coasts of developing countries and in small-island states. This vulnerability may be exacerbated by climate-induced increases in extreme weather events. A 2014 study found that communities in southeast and southwest Alaska, where populations are highly reliant on marine resources and have limited opportunities for alternative employment, face significant socioeconomic risk from ocean acidification. Climate change in marine environments also may threaten human health, with continued ocean warming in tropical and temperate habitats projected to increase the risk of cholera as a result of phyto- and zooplankton blooms and seawater inundation from sea-level rise.[33]

Despite all this, there is good news. The oceans are remarkably resilient.

Conservation efforts aimed at improving system resiliency have proven effective in addressing the nexus between fishing and climate change. For instance, marine protected areas—when effectively enforced—can increase resilience by reducing or eliminating the stress of fishing, allowing ecosystems the extra room to respond to warming temperatures and acidification. Changes in fishing policies to abolish equipment and techniques that destroy benthic (ocean-bottom) habitats and result in bycatch also would reduce the stress of fishing. Finally, revamping our global energy system to dramatically reduce the consumption of fossil fuels would have immense positive impacts on the ocean by curtailing the rise of ocean temperatures and CO_2 levels.

Taking urgent and concerted action to improve ocean health is an imperative, not because saving whales and coral reefs are not worthy pursuits in and of themselves (they are), and not because it is a moral duty (it is), but because—as Melville noted more than 160 years ago—our livelihoods and our lives depend on the sea.

CHAPTER 7

Whose Arctic Is It?

Heather Exner-Pirot

The rapid changes occurring in the Arctic region in the past 10–20 years have become one of the biggest stories in climate change. Temperatures in the Arctic are rising higher than anywhere else on Earth—and more quickly as well. Sea ice has been melting in the summer season at an astonishing rate, and scientists are only beginning to understand the consequences of this thaw for global climate patterns. Many marine species are being affected dramatically by changes in the Arctic environment, with the plight of the polar bear in particular becoming popularized as a symbol of the negative consequences of Arctic warming and global climate change.

In tandem with a warming environment has come growing economic and political interest in the Arctic. Less sea ice ostensibly means more opportunities for shipping and resource extraction, and, troublingly for many, it could result in the opening of previously inaccessible offshore oil and gas fields in the Arctic Ocean and its outlying seas. The "Arctic Paradox"—the irony that global warming related to the burning of fossil fuels will result in new sources of these fuels being extracted in the Arctic—has made the region the most important battleground in the war against climate change.

But that is not the only way to view the Arctic. Although most people in the cities of Europe, North America, and elsewhere seem to see the Arctic through a lens of either climate change or economic opportunity, there is another, often ignored perspective: that of the peoples of the North, for whom the Arctic is not an abstract environmental object but rather a homeland, a workplace, and a community. Many of those who live across the circumpolar north have worked relentlessly over the past 40 years to regain self-determination from national governments and other interests, only to once again feel marginalized by political actors in the mid-latitudes who claim the Arctic as a global commons that is subject to their global governance.

Heather Exner-Pirot is strategist for outreach and indigenous engagement in the College of Nursing at the University of Saskatchewan, Canada, and managing editor of *The Arctic Yearbook*, produced by the Northern Research Forum and the University of the Arctic Thematic Network on Geopolitics and Security.

The Global Arctic and Climate Change

The Arctic often has loomed large in the general public's understanding and perception of climate change. This is due to a combination of environmental and social factors:

Environmental factors. The Arctic, together with the Antarctic Peninsula, has undergone the greatest regional warming on Earth in the past few decades, due to various feedback processes. In the first half of 2010, air temperatures in the Arctic were 4 degrees Celsius (°C) warmer than during the 1968–96 reference period while, over the past half century, much of the Arctic experienced warming of over 2°C, with relative warming increasing at higher latitudes. (See Figure 7–1.)[1]

Figure 7–1. Mean Increase in Global Surface Temperature by Latitude in 2008–2013, Compared to 1951–1980 Baseline

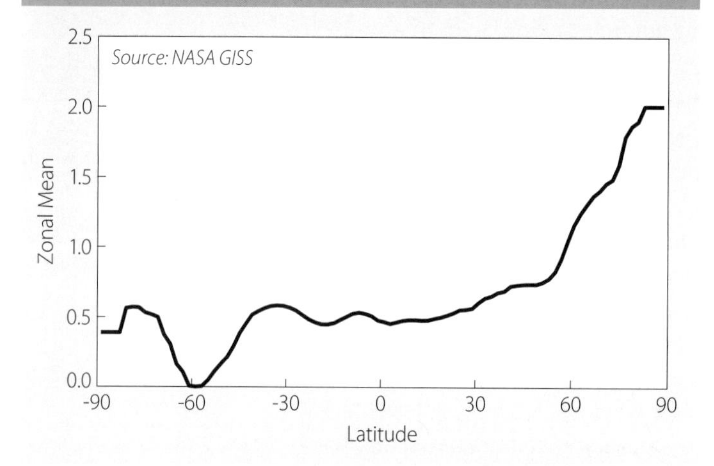

The consequences of Arctic warming are now well known and scientifically documented. Most dramatic has been the loss of summer sea ice, which reached a record low in 2012 of 3.6 million square kilometers, or 52 percent below the 1979–2000 average. (See Figure 7–2.) Overall, summer ice minimum extent, which occurs every year in September, has declined 13.3 percent per decade relative to the 1981–2010 average. Trends are much weaker, although still significant, for the winter ice maximum, occurring every year in March, showing a loss of 2.6 percent per decade.[2]

The loss of sea ice is having a positive feedback effect on the region's climate, creating a situation wherein Arctic warming leads to more Arctic warming. Because white snow and ice strongly reflect solar energy, when sea ice and glaciers shrink, the newly exposed darker waters and lands absorb more solar energy. One study estimates that this absorption is equivalent to as much as one-quarter of the global warming that has resulted from human-caused carbon dioxide (CO_2) emissions.[3]

In another positive feedback mechanism, as the Arctic warms so too does the land-based permafrost that covers large swaths of the region. When permafrost heats up, it releases methane, a powerful, short-lived greenhouse gas that over a 20-year span traps more than 85 times as much heat as CO_2.

Figure 7–2. Average Arctic Sea Ice Extent in September, 1979–2014

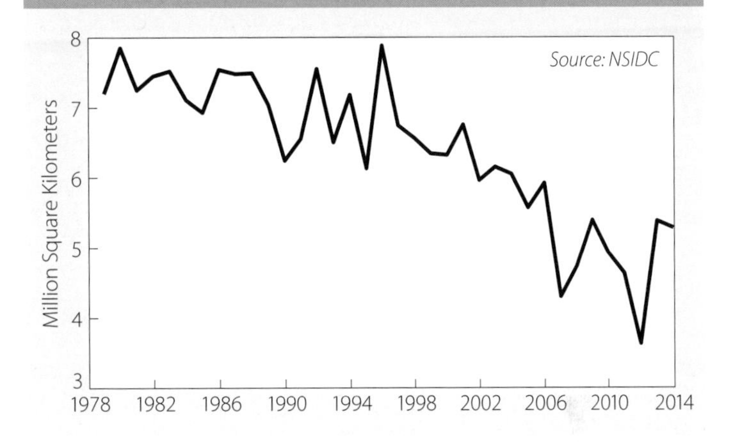

Additional methane is being released in plumes from the thawing Arctic sea bed.[4]

Beyond warming, the increase in CO_2 levels has led to widespread oceanic acidification. Surface ocean waters worldwide are 30 percent more acidic now than at the start of the Industrial Revolution in the late eighteenth century. The Arctic Ocean is especially vulnerable to acidification both because of the large quantities of fresh water that enter the basin (due in part to warming) and because of the water's coldness, which facilitates the transfer of CO_2 from the air into the ocean. Acidification threatens the ocean's shell-building mollusks in particular, weakening their shells and contributing to population declines, which, in turn, affect marine species all the way up the food chain. (See Chapter 6, "The Oceans: Resilience at Risk," for additional discussion.)[5]

Social factors. While the Arctic is being altered observably by climate change, there is a history behind how the region became emblematic of climate change in the popular narrative. First came the science: in 2004, the Arctic Council, the pre-eminent intergovernmental forum for the eight Arctic countries (Canada, Denmark, Finland, Iceland, Norway, Russia, Sweden, and the United States), released its *Arctic Climate Impact Assessment* (ACIA), prepared over five years by an international team of more than 300 scientists, local stakeholders, and other experts. The report presented definitive, scientific evidence of the impacts of climate change in the Arctic and was sanctioned by all eight countries at a time when the issue was still extremely political and contentious.[6]

Leveraging the legitimacy that the ACIA gave to the issue of climate change, in December 2005, Sheila Watt-Cloutier, a Canadian Inuit activist from Kuujjuaq, submitted a petition to the Inter-American Commission on Human Rights in her capacity as chair of the Inuit Circumpolar Council (ICC) claiming that the U.S. government's refusal to limit U.S. greenhouse gas emissions threatened Inuit human rights. Although the petition was not successful, it helped shift public and media attention on climate change from the Antarctic, where the Larsen B ice shelf had famously collapsed in 2002, toward the Arctic region.[7]

Polar Cruises

A young polar bear hopping ice floes.

Former U.S. vice president Al Gore's 2006 documentary, *An Inconvenient Truth*, further entrenched the Arctic as one of the foremost battlegrounds for climate change in the public's mind, with his animated segment of a polar bear struggling to stay afloat amid a lack of ice floes on which to rest and hunt. The impact of a warming Arctic on polar bear populations has become a popular symbol of the need to take action to reduce greenhouse gas emissions.

In the past few years, it has become cliché for politicians, scientists, commentators, and journalists to remark that, "what goes on in the Arctic doesn't stay in the Arctic." Among the most obvious implications of a changing Arctic for the global environment is rising sea levels. As Arctic ice (particularly the land-based Greenland ice sheet) melts, it will contribute to an influx of water to oceans around the world. According to one study, the combined loss from the Greenland and Antarctic ice sheets contributed to sea-level rise of around 1.3 millimeters in 2006, and that rate is accelerating. Since 1900, global average sea level has risen by about 18 centimeters. Hundreds of millions of people live in areas that are prone to flooding, and a majority of the world's big cities are along coasts. A melting Arctic is putting these populations increasingly at risk.[8]

A warming Arctic also may affect weather conditions in the northern hemisphere, as it influences the circulation patterns of the jet stream. There has been some suggestion, based on scientific observations, that the infamous polar vortex—which, in late 2014, brought a harsh winter to much of central and eastern North America as well as to other parts of the northern hemisphere—was linked to climate change and the resulting loss of Arctic sea ice.[9]

The Arctic Region, from the Perspective of the Arctic Region

Considering the dramatic implications of a changing Arctic for the earth's environment, perhaps it is no wonder that many activists in urban and mid-latitude areas have made the Arctic a *cause célèbre*. Perhaps most (in)famous is Greenpeace's "Save the Arctic" campaign, which, according to the group's website, seeks to "defend polar bears," "stop oil spills," and "save our planet" by petitioning world leaders to "create a global sanctuary in the uninhabited area around the North Pole" and to impose "a ban on oil drilling and destructive fishing in Arctic waters." The European Parliament similarly voted in October 2008 to pursue "international negotiations designed to lead to the adoption of an international treaty for the protection of the Arctic, having as its inspiration the Antarctic Treaty."[10]

In general, such promulgations have been anathema to both Arctic states and Arctic peoples. Much of the fault lies in how some southern (i.e., non-Arctic) organizations and politicians have characterized (and caricaturized) the Arctic as a sort of Wild West, where resource exploitation occurs without regulation and oversight, where local populations are defenseless victims of climate changes, and where the region is in need of "saving" by external actors.

ICC Chair Okalik Eegeesiak, a Canadian Inuit from Iqaluit, summed up the situation eloquently at the Arctic Circle conference in Reykjavik, Iceland, in November 2014:

> For whatever reason, many newcomers to the Arctic often see it as a governance vacuum or a region that should be considered a common heritage of mankind. These perceptions overlook the people who live in the Arctic and minimize the importance of existing governance systems. . . .
>
> When I come to these conferences and hear all the plans that people from other parts of the world have for the Arctic, I sometimes feel a bit nervous. . . . We ask that you consult with us before you try to reinvent the Arctic according to your own interests. If you want to help the Arctic, I encourage you to think about what you need to do differently in the South . . . rather than suggesting how to govern ourselves differently in the North. Consider how your activities in the South are impacting us in the Arctic and make some adjustments closer to home.[11]

Consider the irony, from a northern perspective, of southerners beseeching local Arctic communities and governments to make lifestyle changes and to apply bans and moratoriums on drilling, resource extraction, shipping, and fishing in order to reduce the impacts of climate change—impacts that have arisen almost entirely because of activities in southern locales. Northerners are being asked to disproportionately bear the burden of mitigating climate change, even as they disproportionately bear the burden of adapting to those changes.

There is no doubt that northerners have the greatest interest and stake in good environmental stewardship of the region. But it is curious how the Arctic, unlike other inhabited regions of the world, has become a candidate to be a "common heritage of mankind," rather than a political region managed through the same basic processes and governance principles that apply to every other inhabited region in the world—principles that attempt to balance economic, environmental, and social priorities. This is particularly remarkable given that the existing environmental management in the Arctic is as good as, or arguably better than, that in any other region on Earth.

Current Arctic Governance Mechanisms

A variety of mechanisms are currently in place to govern activities in the Arctic Ocean as well as in the various lands of the Arctic region.

Arctic Ocean jurisdiction. The Arctic Ocean basin is juridically divided almost entirely among the five Arctic countries that have shorelines on the Arctic Ocean: Canada, Denmark, Norway, Russia, and the United States (Alaska). Finland and Sweden have no Arctic coasts, and Iceland is considered to be located in the North Atlantic. The United Nations (UN) Convention on the Law of the Sea generally grants countries sovereignty over their territorial waters, which extend 12 nautical miles (22.2 kilometers) from their respective baselines (average low water mark). It also gives special rights regarding exploration and use of marine resources in the Exclusive Economic Zone (EEZ), which extends 200 nautical miles (370 kilometers) from baseline. In cases where the continental shelf extends beyond the EEZ, coastal countries may claim seabed rights to an even wider area.[12]

In practice, this means that it is entirely acceptable under current international law for Arctic countries to drill or fish in their respective EEZs, far into the Arctic Ocean. And if the Commission on the Limits of the Continental Shelf* accepts countries' current bids to extend (or "prolong") their continental shelf areas, more than 90 percent of the Arctic Ocean would likely fall under some level of national jurisdiction under existing international law. (See Figure 7–3.) International shipping access, however, is unaffected in the EEZ except in internal or territorial waters.[13]

It is extremely unlikely that countries would give up accepted rights to Arctic territory that is already theirs under current international law, and choose instead to adopt an Antarctic Treaty-like regime that shares governance with countries or other actors from outside the Arctic region. This does not mean that national and international legal mechanisms governing the Arctic Ocean basin do not exist—and, indeed, several new ones are under consideration. But it does mean that calls for shared governance arrangements and/or the relinquishing of existing sovereign rights are

* The Commission on the Limits of the Continental Shelf made its recommendations on the Norwegian submission in 2009. Canada has yet to submit a full submission but has indicated that it will claim the area up to and including the North Pole. Denmark made its submission on December 15, 2014, and Russia is awaiting recommendations on resubmitting its incomplete 2001 submission. The United States is not party to the UN Law of the Sea Convention (UNCLOS).

Figure 7–3. Prolongation of the Continental Shelf in the Arctic

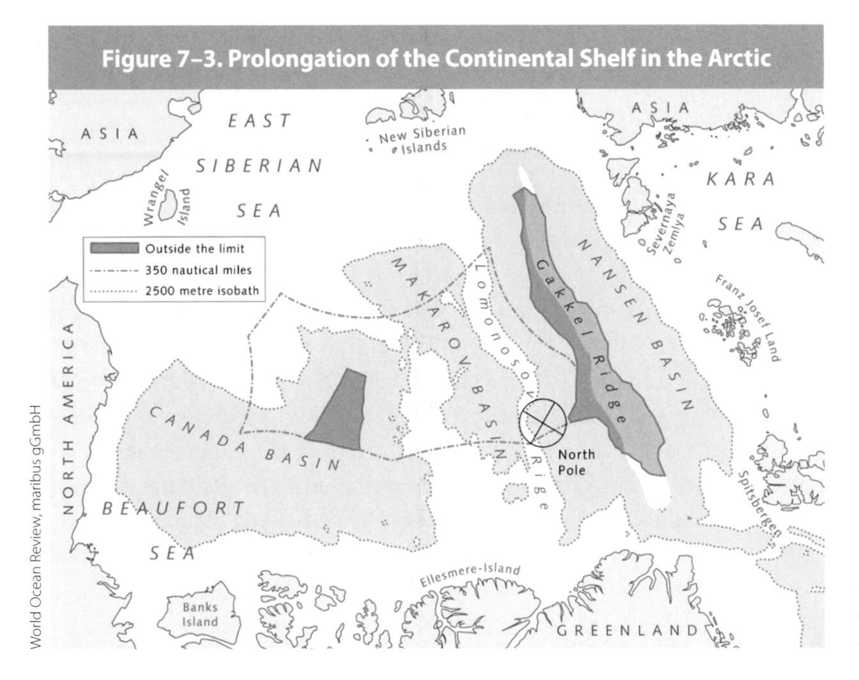

World Ocean Review, maribus gGmbH

Only the areas marked in dark gray are unlikely to be claimable by Arctic countries.

unrealistic and therefore not constructive, and the five countries with Arctic Ocean shorelines have rejected such calls.

Jurisdiction over the lands of the Arctic region. It goes without saying that the eight Arctic countries—the five with Arctic Ocean shorelines, plus Finland, Iceland, and Sweden—have sovereignty over the lands within their national boundaries. It should be noted, however, that the Arctic region is home to many innovative governance arrangements that have granted northern sub-national entities and indigenous groups special rights and particular levels of self-governance. These include:

- The Alaska Native Claims Settlement Act of 1972, which transferred approximately 40 million hectares of public land to native Alaskans, along with a $963 million settlement;
- In Greenland, the attainment of Home Rule in 1979 and then Self-Rule in 2009, which transferred control over a wide variety of governance functions from Denmark to Greenland, including the right to revenues from non-renewable resource development;
- The establishment in Canada of the territory of Nunavut in 1999 and the settlement of land claims in the four Canadian Inuit regions of Nunavik (1975), Inuvialuit (1984), Nunavut (1993), and Nunatsiavut (2005); and
- The negotiation in Canada of the Yukon First Nations Land Claims Settlement Act in 1994, as well as the devolution of additional governance functions to Yukon in 1993 and 2001 and to the Northwest Territories in 2013.[14]

The larger point is that the Arctic region has been undergoing a process of devolution of lands and governing power to regional northern polities—particularly those of indigenous origins—for five decades. The decentralization of political processes and acknowledgement of the indigenous right to self-determination has faced many challenges, but it is widely accepted as the best pathway toward improving quality of life for northern inhabitants.

Arguments originating in southern locales in favor of reinventing the Arctic as a global commons, governed by global interests, are in stark contrast to these efforts to restore governance powers to local communities. It is similarly simplistic to call on Arctic countries to impose bans or moratoriums or to develop laws that may infringe or conflict with the rights that have been granted, sometimes constitutionally, to northern and indigenous inhabitants within their national boundaries. Environmentalists and climate advocacy groups should be reassured by the fact that many of these land claims and governance arrangements include mandatory environmental impact assessments and regulatory processes that are generally as robust, or more so, than standard national regulatory procedures.

Regional and International Environmental Governance

One of the more pervasive myths about the Arctic Ocean is that it is ungoverned. While it is true that some of the common governance structures that are in place in other, more accessible, regional seas are absent in the Arctic, this is largely because the vast majority of the Arctic has been inaccessible to human activity until very recently (aside from subsistence use), and so regulation was moot. Commercial fishing in the Arctic Ocean, for example, has been largely hypothetical until very recently, and shipping is still very limited in Arctic waters.

Because the Arctic region is mostly under the jurisdiction of countries, existing international law applies in the region. Among the major treaties that apply to the Arctic are: the UN Convention on the Law of the Sea*, the Basel Convention on the Control of Transboundary Movement of Hazardous Wastes and Their Disposal, the UN Framework Convention on Climate Change, the UN Convention on Biological Diversity, a broad range of conventions and other instruments adopted by the International Maritime Organization (IMO), the London (Dumping) Convention of 1972 and its 1996 Protocol, the Convention on International Trade in Endangered Species of Wild Fauna and Flora, the Stockholm Convention on Persistent Organic Pollutants, and the Ramsar Convention on Wetlands of International Importance.

---

* Although the United States is not party to UNCLOS, it generally abides by the principles of the Law of the Sea, most of which is customary international law.

Patrick Kelley, U.S. Coast Guard

U.S. and Canadian Coast Guard ships take part in a cooperative survey to help define the Arctic continental shelf.

Relevant non-binding instruments that apply to the Arctic include the Declaration of Principles and *Agenda 21* adopted by the 1992 UN Conference on Environment and Development in Rio de Janeiro, the Global Programme of Action for the Protection of the Marine Environment from Land-based Activities, and the 2002 World Summit on Sustainable Development and its Johannesburg Plan of Implementation. Some regional conventions also are relevant, including the Convention on the Protection of the North-East Atlantic and the Convention on Future Multilateral Co-operation in the North East Atlantic Fisheries, both of which extend into the Arctic region.

In addition, there are Arctic-specific frameworks. In 1991, the eight Arctic countries established the Arctic Environmental Protection Strategy (AEPS) to jointly deal with monitoring and assessment of contaminants, protection of the marine environment, emergency preparedness and response, and conservation of flora and fauna. Although criticized for failing to establish legally binding regulations, the strategy provided for joint cooperation on environmental issues and was very significant politically in the wake of the Cold War. In 1996, the AEPS came under the aegis of the newly established Arctic Council, an intergovernmental forum comprising the eight Arctic countries and including three indigenous organizations (later expanding to six) as Permanent Participants. The Arctic Council was mandated to address issues of sustainable development and environmental cooperation in the region.

The Arctic Council's six working groups have produced exemplary scientific reports on Arctic environmental matters, including the ACIA, the *Arctic Biodiversity Assessment* and monitoring program, an *Arctic Ocean Review* that identifies how management of the Arctic marine environment can be strengthened, and the *Arctic Marine Shipping Assessment.*

The Arctic Council does not have a legal character, and, as such, its recommendations are not legally binding, leading to criticism that it is ineffective. The Council's eight member countries, however, recently have begun negotiating legally binding agreements under its auspices. The first, in 2011, focused on search and rescue. The second, in 2013, addressed cooperation on marine oil pollution, preparedness, and response in the Arctic. It is widely expected that the eight countries will sign additional agreements in 2015, on preventing oil pollution and reducing black carbon and methane emissions, respectively.

That said, the rapidly changing Arctic does demand new governance processes to effectively manage and protect the region's particularly sensitive ecosystem. Many such arrangements are in development. Most prominently, the IMO has been negotiating a mandatory international code of safety for ships operating in polar waters (the Polar Code), to cover the full range of design, construction, equipment, operational, training, search-and-rescue, and environmental protection matters for ships operating in the polar regions. New environmental measures that will come into force once the Polar Code is ratified, likely in 2017, will ban both garbage dumping and oily discharges from ships in polar waters, despite protests from Russia that this would compromise development of the Northern Sea Route. In addition, new voyage-planning regulations will oblige ships to consider whale migration corridors and feeding and breeding areas.

Work is also being done on fish stock management, with the five countries with Arctic Ocean coastlines agreeing in February 2014 to work toward an agreement to block commercial fishing in the ocean's central portion, following the precautionary principle, until more is known about fish populations in the area. No commercial fisheries have operated in the area thus far. Current regional fisheries management organizations that have been established in the marginal seas of the Arctic Ocean include the Northwest Atlantic Fisheries Organization and the North-East Atlantic Fisheries Commission.

Elsewhere in the Arctic, the United States signed a precautionary fisheries management plan in 2009 that prohibited commercial fishing on its side of the Beaufort Sea until scientific research and management measures can ensure a sustainable catch. And Canada announced in October 2014 the establishment of a Beaufort Sea Integrated Fisheries Management Framework for its side. The United States and Russia also signed a bilateral agreement in 1988 addressing fisheries management in the Bering Sea and the North Pacific, although regulation in that region could be more robust.

The next obvious candidate for a legally binding agreement to manage the Arctic region is a Regional Seas Agreement (RSA), which could provide the necessary framework for more consistent and holistic management of the Arctic Ocean. Much of the groundwork toward such an agreement has already been done by Arctic Council working groups, in particular the group on Protection of the Arctic Marine Environment. Admiral Robert Papp, the recently appointed U.S. Special Representative to the Arctic, openly suggested at a conference in Washington, D.C., in September 2014 that the United States would seek to advance an RSA management model while it holds the chairmanship of the Arctic Council in 2015–17.[15]

Economic Opportunities of a Warming Arctic

At least part of the reason that environmentalists and climate advocates have focused on the Arctic region is the unsettling prospect that Arctic warming, which has occurred largely as a result of the burning of fossil fuels worldwide, could result in the exploitation of additional, newly accessible, fossil fuels. This situation has been termed the "Arctic Paradox." Drilling for oil in the Arctic is an almost universally detested concept, probably made worse by the high profile of the *Exxon Valdez* oil spill in Alaska in 1989.

Yet the portrayal of the Arctic as undergoing a "scramble" or "race" for resource exploitation is almost entirely overblown. Even today, the Arctic remains an extremely expensive arena to operate in: it is high-cost and high-risk. A reduction in sea ice may make some parts of the region more accessible, but it also will increase the amount of unpredictable ice floes. And the perpetually dark winters mean that the sea ice will always return for 6–10 months of the year. Costs of construction, maintenance, labor, and exportation of goods are higher in the Arctic than almost anywhere else, and the necessary precautions addressing safety and spill prevention add more costs. There are very high regulatory burdens in the region.

James Brooks

The oil drilling ship *Noble Discoverer* docked in Seattle before traveling to Alaska for the Arctic summer drilling season, 2012.

A handful of ambitious companies has explored Arctic waters, but with little success. Shell has spent eight years and $6 billion exploring in the Alaskan Arctic, but this effort has been plagued by a series of setbacks, prompting ConocoPhillips and Statoil to suspend their Alaskan Arctic drilling plans. Cairn Energy has spent $1.9 billion drilling eight test wells off the northwest coast of Greenland, but the company announced in January 2014 that it would not conduct further drilling operations that year. Meanwhile, operations in the Shtokman field off the coast of northern Norway and northwestern Russia have been suspended indefinitely due to low gas prices resulting from the global shale gas glut, despite capital costs estimated at $15–$20 billion absorbed by investors from Gazprom, Total, and Statoil. All of these serve as warning signs, preventing new investments in Arctic oil drilling in the short and medium term.[16]

In the meantime, it is worth asking whether a ban on oil drilling would be ethical or even legal. Northern populations are sparse, workers are generally unskilled, and distances to markets are great. The best—and perhaps only—opportunity that northern communities have for development is resource exploitation, whether mining, fishing, or hydrocarbons. Southern activists inevitably conjure up the Shells of the world when they think about Arctic economic development, but for many communities and governments, poverty reduction is foremost in their minds.

Consider the case of Greenland, which for decades has sought greater independence from its former colonial master, Denmark, and has succeeded in obtaining a significant measure of self-determination. But Greenland will continue to depend on Denmark for an approximately $640 million annual subsidy for its 58,000 residents until it can replace these funds with an equivalent source—namely, resource revenues. In that sense, those who advocate for a ban on oil drilling in the Arctic are condemning Greenlandic Inuit to continued dependence on a European nation, or at the very least removing the ability of Greenlanders to decide for themselves if the environmental costs of drilling are worth the social and economic benefit. The right to weigh environmental, economic, and social costs when making governance decisions is enjoyed by all other sovereign nations and should not be expected to be relinquished by Arctic states and peoples to address the consequences of actions committed elsewhere.

Social and Economic Sustainability of the Arctic

It is important to recall that sustainability is not merely an environmental concern. As the Brundtland Report, *Our Common Future*, famously articulated in 1987, "Sustainable development must meet the needs of the present . . . in particular the needs of the world's poor to which overriding priority should be given. . . . The satisfaction of human needs and aspirations is the major objective of development."[17]

Incredible progress has been made in the circumpolar north to restore self-determination to northerners and in particular to indigenous peoples who may have very different values and goals than those found in the political centers of the eight Arctic countries. Setting a context in which local groups can be partners in resource development (or have the right to limit such development within their territory), and can make cost-benefit analyses of the jobs and public revenues that such development brings, has been a great political achievement of the past 40 years. Seeing the Arctic exclusively as an ecosystem in need of preservation—and not as a homeland where people have a right to live and work and improve their standard of living through economic development—imposes a hidden threat to the long-term sustainability of the region by removing the ability of northerners to make decisions about their lands and territories.

This does not mean that development should, or will, happen at any cost. It means that we should be careful about applying a standard of environmental protection in other regions that we would not accept in our own. Perhaps where international advocacy is needed is in those northern regions where local inhabitants do not have a say in how development proceeds. Support should go toward advancing local agendas in the north, not leveraging them to advance agendas in the south.

Conclusion

There is no doubt that the Arctic is undergoing significant and potentially devastating changes as a result of climate change. The seriousness of these changes should compel governments and individuals to act. Unfortunately, much of the recent focus in mitigating climate change is centered on the Arctic itself, despite the fact that the people living there are responsible for only a miniscule share of the human-caused greenhouse gas emissions that have precipitated climatic changes in the first place. Many southern environmentalists erroneously conflate Arctic impacts with Arctic activities, and develop their strategies accordingly. It would be far more constructive for them to work on reducing fossil fuel use in their own regions, rather than seeking to manage the consequences of this energy use in others.

For their part, the governments of the eight Arctic countries have made impressive and discernible advances in protecting the Arctic environment in the context of rapid change, including by addressing the impacts of externally induced climate change through the Arctic Council. Critics may rightly point out that the Council's actions are not legally binding, that it is often under-resourced, and that it takes a long time to make decisions. But that is only compared to the ideal. When compared to any other regional or international intergovernmental forum, the Arctic Council is as progressive, action-oriented, and efficient as they come. This is made all the more impressive by the fact that this body includes both Russia and the United States, countries that are ideologically at odds on many issues.

From a northern perspective, climate change is having real consequences for traditional ways of life, food systems, infrastructure, and external relations. Indigenous peoples and other northerners have the greatest stake in protecting the Arctic environment—their home. As such, it is unconscionable for southern activists to attempt to deny Arctic peoples their right to make decisions on the region's governance by calling for global bans or conservation areas to reduce economic activity that might contribute to climate change, when no other society in the world has accepted such restrictions. The eight Arctic countries should certainly do more to reduce their carbon emissions, but it is far from obvious that the Arctic Council is the best forum in which to do so. International, rather than regional, frameworks—beginning with the UN Framework Convention on Climate Change—are much better placed to discuss and negotiate carbon reductions, which then must be implemented at a national level.

Despite the fact that the Arctic region is not the source of global climate change, it has become necessary to address the consequences of global warming there. Arctic political actors have responded in kind. We would do well if the other political regions on the planet made as much effort to mitigate and adapt to climate change as the Arctic has, and we should focus our efforts accordingly.

CHAPTER 8

Emerging Diseases from Animals

Catherine C. Machalaba, Elizabeth H. Loh, Peter Daszak, and William B. Karesh

In December 2013, an outbreak of the deadly Ebola virus began in a small village in southern Guinea, the first outbreak of the Zaire Ebola strain in West Africa. Within a year's time, the outbreak, which was not officially noticed by health authorities until March 2014, had led to approximately 18,000 known human cases and 6,300 deaths, posing an unprecedented challenge to global public health. Air travel helped the disease leap from West Africa to other continents, including North America and Europe.[1]

Despite global attention and response, 12 months into the outbreak the initial source of human infection still had not been identified. Prior Ebola outbreaks in humans, as well as a concurrent outbreak in the Democratic Republic of the Congo beginning in August 2014, have been linked to the hunting or handling of wild animals, with subsequent transmission among humans. Certain bat species are the suspected natural source for the virus and are thought to harbor it without signs of disease. Researchers have detected Ebola infection and mortality in wild chimpanzees, gorillas, and duiker antelopes, and evidence from human outbreaks suggests that these species have served as brief hosts for potential human infection when hunted or handled. Studies suggest that Ebola is causing severe declines in great ape populations—particularly critically endangered wild lowland gorillas—making it as much a threat to biodiversity as it is to human health.[2]

Around the same time that the Ebola outbreak was spreading through West Africa, a human case of a different disease, caused by another pathogen in the same family of viruses, emerged in Uganda. The infected patient experienced symptoms including fever, abdominal pain, vomiting, and diarrhea and ultimately died a few weeks into the illness, caused by the Marburg virus. While the source of this particular outbreak was not known, past human cases of Marburg originated from contact with certain species of cave-dwelling bats that serve as the natural carriers of the virus.[3]

Ebola and Marburg viruses are just two examples of an emerging but

Catherine C. Machalaba is program coordinator for health and policy, **Elizabeth H. Loh** is a research scientist, **Peter Daszak** is president, and **William B. Karesh** is executive vice president for health and policy, all at EcoHealth Alliance. This chapter is adapted from William B. Karesh, Andy Dobson, James O. Lloyd-Smith, Juan Lubroth, Matthew A. Dixon, Malcolm Bennett, Stephen Aldrich, Todd Harrington, Pierre Formenty, Elizabeth H. Loh, Catherine C. Machalaba, Mathew Jason Thomas, and David L. Heymann, "Ecology of Zoonoses: Natural and Unnatural Histories," *The Lancet* 380, vol. 9857 (December 1, 2012): 1936–45.

largely overlooked trend: the spread of infectious disease from animals to humans. The emergence of such "zoonoses," responsible for a growing number of disease outbreaks that have sickened or killed millions, is facilitated by the human disruption of natural ecological conditions, which has allowed for increased human-animal contact. Despite the extensive public health response to these emerging infectious diseases, the focus has been on reactive rather than preventive efforts. But new strategies for dealing with these threats offer the possibility that such diseases need not be a threat and a scourge, and that humans once again can learn to live in balance with the natural ecology that supports us.

Pandemics of Animal Origin: A Growing Threat

For millennia, humans have been stricken, sometimes seriously so, by pathogens originating in animals. Many diseases that are commonly known to be transmitted among people, such as measles and (formerly) smallpox, evolved from microbes living in wildlife. And many of history's most devastating pandemics have animal origins, including the Justinian Plague (541–542 AD), the Black Death (Europe, 1347), yellow fever (South America, sixteenth century), and the global flu outbreak of 1918—as well as modern pandemics such as HIV/AIDS, severe acute respiratory syndrome (SARS) in 2003, and the highly pathogenic H5N1 (avian) flu.

Mark A. Wilson

Portal of entry: commemorative plaque in Weymouth, England.

Today, diseases of animal origin account for about two-thirds of human infectious diseases, causing about a billion cases of human illness and millions of deaths each year, and racking up hundreds of billions of dollars in economic damage over the past two decades. Most known zoonoses are "endemic," meaning that they tend to be confined to a particular region. These endemic infections—such as rabies or trypanosomiasis (sleeping sickness, transmitted by the tsetse fly)—typically pass from animals to people with little or no subsequent person-to-person transmission.[4]

But when an endemic zoonosis crosses into a new geographical area or host species, or evolves new traits (such as drug resistance)—or when a novel pathogen is transmitted to humans for the first time and causes an outbreak—it becomes an "emerging" zoonosis. Emerging zoonoses from wildlife account for most of the emerging infectious diseases identified in people in the past 70 years. Their spread typically is facilitated by human activities, including changes in land use, population growth, alterations in behavior or social structure, international travel or trade, microbial adaptation to drug or vaccine use or to a new host species, and breakdown in public health infrastructure. These activities give zoonoses tremendous range: with more than 1 billion international travelers every year, as well as the extensive international trade of wildlife, infected individuals could potentially spread zoonotic diseases anywhere in the world.[5]

In the past few decades, accelerating global changes have led to the emergence of a striking number of newly described zoonoses, including hantavirus pulmonary syndrome (a respiratory disease contracted from infected rodents), monkeypox (similar to smallpox, and transmitted from a variety of animals), SARS (a pneumonia spread by small mammals), and simian immunodeficiency virus (the animal precursor to HIV). Some of these zoonoses, such as HIV, have become established as serious diseases that now pass from person to person without repeated animal-to-human transmission.

Ecology of Disease

Like any infection, zoonoses emerge when a chain of infection is activated—a process whereby the pathogen or infectious agent passes from the reservoir host in which it naturally occurs, or from an intermediate host species, to a susceptible host and is ultimately pathogenic to humans. For infection to occur, all six elements of the chain of infection must be present, from the disease-causing agent, to the mode of transmission, to the susceptible host. (See Box 8–1.) In its simplest form, this chain is straightforward—but any of the elements can present complications.[6]

Consider a case where an animal species, such as a small rodent, can be a reservoir host (carrying the infectious agent), but it also can host ticks (a vector for spread of infection of some pathogens)—thus complicating and potentially increasing the opportunities for dissemination. White-footed mice are a natural reservoir of the bacteria that cause Lyme disease and can spread the bacteria to ticks that feed on the mice, enabling the infection to spread to other species that the ticks feed on, including humans. Some zoonoses can have several reservoirs or intermediate host species, each of which might have a different role in a pathogen's emergence. The Nipah virus, which lives in fruit bat reservoir hosts in Malaysia, also became established in domestic pig populations in the 1990s, amplifying viral transmission and leading to a large human outbreak in 1998–99 that killed 100 people and led to the slaughter of more than a million pigs as a control measure.[7]

Human activities can change the ecologies underlying the chain of infection of zoonoses, such as when these activities alter the size of the host

Box 8–1. The Chain of Infection

Development of an infection has six components:

Agent of disease. The disease-causing organism, or pathogen, which can take the form of a bacteria, virus, fungus, or parasite.

Reservoir. The species—human, animal, or insect—in which the pathogen naturally resides. Pathogens can live in a reservoir for long periods without emerging to cause an epidemic. Reservoir hosts may not be seriously harmed by the pathogen.

Portal of exit. The path by which a pathogen leaves its reservoir or host. Examples include the respiratory tract, urinary tract, rectum, and cuts or lesions in skin.

Mode of transmission. The way a pathogen spreads from its reservoir host to the susceptible host. This can occur directly, via skin-to-skin contact or sexual relations, or through the spread of droplets from coughing or sneezing. It also can occur indirectly, as when organisms are carried on airborne particles, when intermediate objects such as handkerchiefs or bedding are the vehicle of transmission, or when mosquitoes, ticks, and other vectors carry the pathogen.

Portal of entry. The place a pathogen enters a susceptible host. The mouth and nose are common portals of entry. Others include the skin (for hookworm), mucous membranes (for influenza or syphilis), and blood (for hepatitis B and HIV).

Susceptible host. Some host species can acquire the pathogen but do not naturally carry it, and may be affected or unaffected by it, potentially transmitting it to other species or populations or serving as a dead-end for transmission.

Importantly, human activities can facilitate the transmission of a pathogen at any of these six places—by, for example, enabling contact between reservoir and host species or inducing genetic selection for virulent strains that are more likely to be pathogenic to humans. Conversely, human intervention around any of the six components can stop the spread of an infectious disease.

Source: See endnote 6.

population. Reducing the population of a preferred animal host, such as a large, hoofed animal, can cause a transmitter, such as a mosquito, to shift its feeding patterns to humans, leading to a disease outbreak. After cattle imported from Asia introduced a viral disease known as rinderpest, or "cattle plague," to Africa, both cattle and wildebeest populations in Africa declined rapidly and tsetse flies switched to feeding on people, causing a large epidemic of sleeping sickness.[8]

Sometimes, a naturally occurring or environmental change can lead to a change in the size of host populations, increasing the risk of transmission to humans. El Niño events in 1991–92 and 1997–98 led to the appearance of human hantavirus cases in the southwestern United States, via an ecological cascade: increased precipitation caused vegetation growth, which supported increased populations and densities of rodents, which, in turn, facilitated hantavirus infections in rodents. These changes increased the risk of infection in people.[9]

Ecological principles also apply to the dynamics of pathogens within individual hosts. Pathogen populations living within an infected host grow and evolve according to the same competitive principles that govern the growth of plants or animals living freely outside a host. This competition between pathogens and other microbes within a host, in addition to molecular factors and the mode of transmission, can determine how great a threat the pathogen poses to human health. Shifting the diet of beef cattle before slaughter, for example, creates new environmental conditions within the gut of the animal that can increase the population of human pathogens, such as the foodborne bacterium *E coli* that can result in illness and even death.[10]

The community of commensal (or co-existing) bacteria—such as the "good bacteria" in the gut that help with the digestive process—also plays an important part in combating pathogens. Disruption of this community through changes in diet or through the use of antimicrobial remedies can allow the growth of other organisms, some of which might be pathogenic. This disruption may explain some of the increased risk of zoonotic infections for salmonella, for example. The vital role played by commensal bacteria underscores the importance of studying the full microbial community within a host, and not just pathogens.[11]

Livestock and Wild Animals

People eat a wide range of animals, both farm raised and wild, and many of these can harbor bacteria, viruses, or parasites that can be transmitted to humans. This makes the production, processing, and consumption of livestock, as well as the hunting, preparation, and consumption of wild meat, potential paths of disease transmission.[12]

As human societies develop, each era of livestock revolution presents new health challenges and new opportunities for the emergence of zoonotic pathogens. Pathogens found in livestock production processes have caused repeated outbreaks of bovine tuberculosis, brucellosis, salmonellosis, and other illnesses that result from new cultural and farming practices. Livestock production practices that can create challenges for animal health include high stocking rates, mixing of species, prophylactic use of antimicrobials for growth promotion, and poor implementation of disease surveillance and control measures. These practices often are found in areas where the veterinary infrastructure is weak and where the public-private partnerships, policies, and capacities to support and strengthen it are insufficient.[13]

Meanwhile, livestock raising in concentrated feeding operations (or factory farms), a common practice in industrial countries and increasingly in developing countries, may heighten the risk of dissemination of animal diseases to humans. Intensification offers economies of scale, but it also can contribute to the spread of disease by increasing the density of potential host populations, raising contact rates among animals, reducing genetic

EPA

Concentrated feeding of hogs in the United States.

diversity within populations, and prioritizing species that are good at converting feed over those with higher disease resistance. The highly pathogenic H5N1 bird flu, which killed hundreds of people in Asia in the early 2000s, likely evolved into such a virulent strain because of high rates of mixing among flocks, and it spread because of marketing practices and the contamination of bird-raising environments. Hundreds of millions of birds were killed by the flu or had to be killed to prevent its spread.[14]

In addition, methods of slaughtering and processing animals; storing, packing, and transporting products; and preparing foods in the home can facilitate outbreaks of foodborne diseases. Incomplete cooking of pigs and wild boars can lead to trichinosis and cysticercosis, the latter afflicting 50 million people annually (often subsistence farmers in developing countries) and resulting in epilepsy and even death. Echinococcosis, caused by the larval stages of a tapeworm that is transmitted via hoofed animal hosts, is spread through the ingestion of inadequately prepared food, affecting 200,000 people every year and costing more than $4 billion annually for treatment and control. Other notable parasites transmitted through inadequate food processing and preparation include trematodes (liver, lung, and intestinal tapeworms), a neglected disease group that poses a serious threat to public health and economic prosperity in Southeast Asia.[15]

Globally, people consume far fewer wildlife products than they do livestock, but the human demand for wild meat is not inconsequential: in central African countries alone, people eat an estimated 1 million tons of wild meat annually. Human contact with animals through the hunting, preparation, and consumption of wild animals has led to the transmission of deadly diseases such as HIV/AIDS (linked to the butchering of hunted chimpanzee), SARS (which emerged in wildlife markets and among restaurant workers in southern China), and Ebola. In each case, the organisms or pathogens exploited new opportunities that resulted from changes in human behavior.[16]

Land-Use Change

Large-scale changes in land use contribute to the spread of many zoonoses, by affecting biodiversity and the relations between animal reservoirs and other animal hosts or vectors, people, and pathogens. Land modification can lead to changes in vegetation patterns, microclimates, human contact with animals (both domestic and wild), and the abundance, distribution, and demographics of vector and host species, all of which are critical factors in disease ecology.

In the region surrounding the town of Lyme, Connecticut, a repeated cycle of deforestation, reforestation, and habitat fragmentation changed the dynamics of predator-prey populations and led to the emergence and spread of Lyme disease, now the most common vectorborne illness in the United States. The mobility of ticks and other carriers has enabled the disease's observed northward and westward spread over the past decade. Similarly, the origin of human alveolar echinococcosis, a disease associated with a tapeworm that often resides in small mammals (especially rodents), has been traced to Tibet, where overgrazing and degradation of pastures increased the population densities of small mammals, which served as intermediate hosts for the disease and passed it to humans.[17]

Many tropical regions are emerging disease hotspots, rich in diversity of both wildlife and microbes—many of which have not yet been encountered by people. The opening up of tropical forests for plantation development and extractive industries such as mining, logging, and oil and gas may increase the risk of zoonotic disease by changing the composition of habitats and vector communities, altering the distribution of wild and domestic animal populations, and increasing exposure to pathogens through increased human-animal contact. Among the infectious diseases associated with changes in tropical land use are Chagas disease, leishmaniasis, and yellow fever—all of which are life-threatening illnesses spread via infected insects.[18]

Human contact with wildlife is increasing on a large scale through road building, the establishment of settlements, and the rising mobility of people, as well as through the extractive processes themselves. In areas where such changes take place, the hunting, consumption, and trade of wildlife for food often rises. If a site is poorly managed, the growing human population can strain existing infrastructure, leading to overcrowding, poor sanitary conditions, improper waste disposal, and a lack of potable water. All of these changes increase the risk of cross-species transmission of pathogens, resulting in zoonotic disease. Recent human immigrants to an area may not have immunity to zoonotic diseases that are endemic to that area, making them particularly susceptible to infection.[19]

Although extractive industry companies often do assessments of the environmental and social impacts of their activities, these studies rarely include principles of disease ecology because standard operating procedures in developing countries and specific laws or regulations often do not require the assessment of health risks at a community level. And although some assessments do include zoonotic disease from domestic animals in their guidelines, few adequately address the full range of potential zoonotic pathogens, especially from wildlife.[20]

Resistance to Antimicrobial Drugs

Injudicious use of antibiotics and other antimicrobial remedies in animals can leave people vulnerable to the spread of infectious disease. The most direct mechanism for the evolution of antimicrobial-resistant infectious diseases in people is the use of antibiotics in treating human infections. But the widespread use of antimicrobial drugs in livestock production—both to prevent disease and to promote animal growth—has led to worries about this being another possible route for emerging antibiotic resistance in people. Not only may genetic selection pressures from antimicrobial use lead to development of resistant strains, potentially posing food security risks and zoonotic disease risks for livestock handlers, but antimicrobial exposure may also occur via the food chain as well as through environmental dispersion (e.g., through manure, runoff, etc.).[21]

From an ecological perspective, antimicrobial resistance is a natural occurrence. Genes conferring resistance probably originated as an evolutionary response to antimicrobial compounds that bacteria, fungi, and plants living freely in the environment produced to protect themselves from infection or competition. The early antibiotics used in human medicine all were derived from natural bacterial and fungal sources. Over time, use of these compounds resulted in selection for resistance in bacteria, and horizontal transfer allowed these genes to spread rapidly through microbial populations and communities. Today, antimicrobial resistance is emerging based on these same evolutionary principles, with microbial populations adapting through competition and selection. But because the use of antimicrobial agents in people is far more widespread now than it was when these drugs were developed, the potential for emergence of resistance is likely much more rapid.[22]

The common practice of administering antimicrobials to livestock may be contributing to this trend. Increased intensification of livestock production over the past half century has created dense host populations that readily transmit disease. In response, agricultural industries introduced a range of antimicrobial drugs to combat the spread of infection among closely confined animals. In addition to being used prophylactically, some of these antibiotics are used in animal feed to enhance growth rates, improve feeding efficiencies, and decrease the animals' waste output.[23]

The question of whether antibiotic use in agriculture has exacerbated drug resistance in people is widely debated. Farm workers who were exposed to antibiotics through their jobs showed an increased prevalence of resistant bacteria in their gut, and studies have reported instances of farm animals containing resistant pathogens of relevance to human medicine—including a strain of *Staphylococcus aureus* that is resistant to methicillin, a first-line antibiotic once commonly used to prevent Staph infections. It is possible, however, that these bacteria were passed to the animals from people.

Antimicrobial-resistant pathogens may be transmitted from livestock to people in several ways, including food consumption, direct contact with treated animals, waste management, use of manure as fertilizer, fecal contamination of runoff, and relocation or migration of animals. Additionally, some 30–90 percent of veterinary antibiotics are excreted after being administered to livestock—mostly in an unmetabolized form—providing a route for dissemination and potentially exposure in the environment.[24]

Combating Zoonoses

The recent re-emergence of Ebola in the Democratic Republic of the Congo, as well as the ongoing challenge of the HIV/AIDS pandemic, are sober reminders of the serious threat that zoonotic pathogens pose to human well-being. These global health challenges are also a reminder that traditional approaches to identifying potential new human pathogens—such as tracing back to the host source of a human disease once it has emerged—may be of limited effectiveness in preventing ongoing human transmission. (Such approaches probably would not, for example, have identified simian immunodeficiency virus, the forerunner to HIV/AIDS, as a potential risk to humans.) Thus, bold new approaches to the prevention of zoonoses are needed.[25]

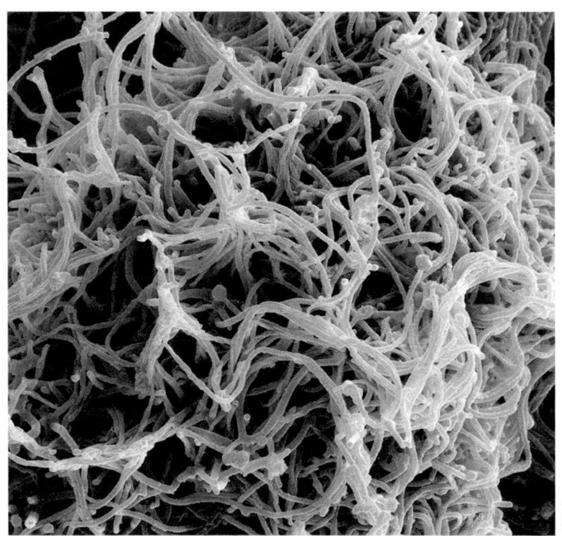

NIAID

Scanning electron micrograph of filamentous Ebola virus particles budding from an infected cell.

Understanding the ecology of zoonotic diseases is a complex challenge. It requires knowledge of animal and human medicine, ecology, sociology, microbial ecology, and evolution, as well as of the underlying dynamics that increase the transmission of pathogens in humans, wildlife, and livestock. The so-called One Health perspective, which considers this wider web of interactions and dynamics, incorporates a critical understanding of how the environment is changing, and how these changes, in turn, affect microbial dynamics. Because of the wide range of disciplines involved, preventing and responding to zoonotic diseases requires a multidisciplinary effort, with collaboration among ministries of health, environment, and agriculture; within and across governments; and with intergovernmental agencies involved in health, trade, food production, and the environment.[26]

As one key to a multisectoral approach to zoonosis prevention, ecologists and clinicians need to collaborate in early-detection and control programs. Combining ecological science and real-time clinical data could improve the accuracy of mathematical models, the design of prospective and retrospective studies, and the outcomes of field studies seeking to identify key risk factors. In addition, great value would accrue if public health scientists (who use epidemiological techniques and rely on human case data) collaborated closely with disease ecologists (who often work with wildlife or livestock data) to model risk in human beings. Such disease ecology approaches might be useful not only in containing an established outbreak, but also in predicting the emergence and spread of new zoonoses. Understanding the relationship between environmental changes; the dynamics of wildlife, domestic animal, and human populations; and the dynamics of their microbes can be used to forecast the risk of human infection from zoonoses.[27]

Frequently, the dynamics of pathogens in the non-human reservoirs of a zoonosis (apes, mosquitoes, mice, etc.) determine the risk of outbreak in people. This risk can vary with geography, the season, or across multiyear cycles, and is influenced by changes in land use, weather, climate, and the environment. Knowing the dynamics of zoonotic pathogens in their wildlife reservoirs could help in creating an early-warning system to alert authorities of the risk of an outbreak in livestock or people. In the case of Rift Valley fever, the density of vegetation correlates with breeding sites for the mosquito vectors, and satellite monitoring of this density has been used to forecast cases of the disease in people and to predict the need for vaccines. Such approaches can be refined and developed, and eventually used to predict the risk of future disease emergence.[28]

Other ways to further global disease prevention capacity and efforts include implementing the World Health Organization's International Health Regulations, which make it easier to report a broad range of human disease events, and supporting implementation of the World Organisation for Animal Health's international standards for animal health, which require the reporting of animal diseases, including zoonoses. Improving veterinary services in many low-income and middle-income countries can help to expand awareness of zoonotic diseases, the ability to detect and prevent them in animals (including wildlife), and the ability to quantify and report their occurrences. Because of the high economic costs of zoonotic diseases to both commerce and society, it could prove more cost effective to try to prevent and control these diseases by integrating science-based control strategies in animals, rather than seeking to control the illnesses in people alone.[29]

Because approximately three-quarters of recently emerging diseases in humans have originated in wildlife, an early step in preventive efforts should be to identify the diverse pathogens that wildlife harbor, as well as the characteristics that make them risks to human health. Researchers estimate that detecting 85 percent of the viral diversity in mammals would cost around $1.4 billion, or $140 million per year over 10 years. This is a small fraction of the cost of an emerging disease event (the 2003 SARS outbreak, for example, cost the global economy an estimated $30 billion-plus). The public health community can use the information gained from this effort to better identify emerging disease threats and to take measures to prevent outbreaks in both humans and other species that we depend on for nutrition, other resources, and ecosystem functions. Routine disease surveillance of animals also may help with early detection of health risks to humans.[30]

Sarahtz

Selling bush meat at Makenene market, Cameroon.

New avenues of research are needed to understand the complex ecology of antimicrobial resistance and foodborne zoonoses, including how the microbiomes of both humans and the animals that we interact with work, and what causes zoonotic microbes to proliferate. The effects of antibiotic use in livestock are not well understood, but involving physicians, veterinarians, and ecologists in the design and interpretation of studies could advance our understanding of this area. Standardized data collection, long-term monitoring, and risk assessments are needed to better understand the development of multidrug resistance and multibacterial infections, from the use of antimicrobials in livestock as well as from wildlife. To reduce the need for antimicrobial use in people and animals, alternatives such as probiotics, diets to promote healthy or protective gastro-intestinal flora, and new methods of immune system modulation need to be explored.[31]

Extractive industries, such as mining and oil production, can be part of disease prevention as well, by helping to minimize the opportunities that enable transmission of pathogens that are new to human hosts. Guidelines are needed urgently for safe or best practices that include ecological knowledge to reduce the risk of disease emergence or occurrence. Disease risk analysis tools can be used to determine the potential health impacts of ecological change from potential human activities, allowing for proactive interventions that will mitigate risks. For example, industries establishing work sites (such as mining operations) in remote areas could be required to provide food sources for their employees to reduce subsistence hunting of wildlife. Such guidelines could be required by development banks or other public agencies that finance large-scale projects, or by insurers.

The wide gaps that exist between industrialized and developing countries in public health, veterinary, and medical infrastructure and training affect efforts to prevent, monitor, and control disease. In addition, ecological approaches for preventing and controlling zoonotic disease are not used in most countries. These challenges need urgent attention, and the One Health approach provides a promising holistic framework for achieving this aim.

Although the causes and risks of zoonoses vary widely across regions and cultures, increasing global connectedness demands the attention and alertness of health professionals everywhere. Because human activities are a driving force for where and how zoonoses occur, not only are improved healthcare systems needed, but multisectoral approaches to mediate the impact of human activities on disease dynamics are indispensable to contain zoonoses and prevent the emergence of new ones.

CHAPTER 9

Migration as a Climate Adaptation Strategy

François Gemenne

People react to environmental degradation in many diverse ways. It has long been recognized, however, that changes to the environment can induce significant population movements, either as a direct consequence of these changes or because of the impacts that environmental changes have on other drivers of migration, such as poverty or food security. In recent years, scholars and policy makers alike have expressed rising concern that climate change could become a key driver of migration in the coming decades.[1]

Already in 1990, the Intergovernmental Panel on Climate Change warned that "one of the gravest effects of climate change might be that on human migration." An ever-growing body of literature has addressed this issue, but most of the work has focused on the number of people who could be displaced, the influence of environmental factors in the decision to migrate, or the legal and humanitarian challenges posed by the projected new flows of migrants. As researchers Jon Barnett and Michael Webber have noted, reports on the topic "rarely recognize the potential for spontaneous and planned adaptations to reduce vulnerability to environmental change," nor do they "adequately recognize that migration is itself a strategy to sustain livelihoods."[2]

Most of the literature on the subject has been widely alarmist, with some reports citing made-up projections of hundreds of millions of "climate refugees" worldwide by 2050. Climate-induced migration has been presented variously as one of the most dramatic consequences of global warming, as a humanitarian catastrophe in the making, and as a threat to international security. Yet long-time research on livelihoods and adaptive capacity makes clear that populations affected by environmental changes frequently have used migration as a deliberate adaptation strategy, especially in Africa's Sahel region. Until recently, this body of literature was largely overlooked by the dominant research on climate-induced migration, which consistently presented migration as a failure to adapt to environmental changes.[3]

François Gemenne is executive director of the Politics of the Earth program at Sciences Po in Paris and a senior research associate with the University of Liège in Belgium.

Despite the variety of claims concerning climate-induced migration, the empirical reality is quite different. Although climate change can induce dramatic population displacement, the common conception tends to present migrants solely as resourceless victims. The overall number of people that may be compelled to move as a result of climate change is as-yet unknowable, but it is likely to be large, and this migration will likely involve mass suffering. The character of that suffering could take many forms, including responding to the evolving situation in place, migrating within one's country, and possibly migrating across borders.

These distinctions matter because they affect the type of policy responses needed. The conception of migrants solely as victims, however, might actually hinder their capacity to adapt, and induce inadequate policy responses. Fortunately, some policy directions are available that would allow migration to unleash its adaptation potential.

Common (Mis)perceptions of Climate-Induced Migration

Migration often is perceived as a decision of last resort that people take when they are faced with environmental disruptions. It is commonly assumed that migrants have exhausted all possible options for adaptation in their place of origin, and are left with no other choice but to flee. Reports on the impacts of climate change are replete with the idea that climate-induced migration should be avoided at all cost, and would represent a failure of policies designed to help populations mitigate and adapt to climate change. Only a few analyses have considered that migration actually could be a resource that migrants use to deal with environmental changes. The result of this misconception is that migrants usually are perceived as resourceless, expiatory victims of climate change.[4]

Erin Magee, Australian Department of Foreign Affairs and Trade

Most Kiribati government buildings are on South Tarawa.

Over time, "climate refugees" have become the human face of global warming, simultaneously being the first witnesses and the first victims of climate impacts such as sea-level rise or melting permafrost. Populations in low-lying island nations, such as Kiribati or the Maldives, are portrayed as the proverbial canaries in the coal mine—alerting the rest of the world to the dangers of climate change, and themselves left with no choice but to relocate abroad. Strikingly, many of these populations refuse to be considered as refugees in the making, as it would undermine their adaptive capacity and render ineffective the efforts they already have undertaken to adapt.[5]

The perception of migrants as victims is deeply rooted in environmental determinism, a perspective that asserts that an individual's course of action is determined exclusively by his or her environment. An all-too-common view is that most people affected by environmental changes would need to migrate, and that environmental factors would be the sole drivers of their migration. Examples of this perspective can be found in numerous reports on climate impacts (and on sea-level rise in particular), which also attempt to forecast the number of people who potentially could be displaced. The consequence of this deterministic perspective is that "environmental migration" often has been viewed as a new and distinct migration category, the nature and magnitude of which would be determined by environmental changes only, and where migrants would be set apart from broader global migration dynamics.[6]

From a policy point of view, analysts have associated this new category of migrants with specific policy challenges. A common assumption has been that massive flows of migrants from poor countries soon would be flocking to the doors of industrialized countries. An image on display at the Museum of London's 2010 exhibition "Postcards from the Future," for example, showed Buckingham Palace surrounded by a shanty town of "climate refugees"—just one vision of what the city's landmarks could experience in an environment transformed by climate change.[7]

Policy-wise, climate-induced migrations also have been presented as an impending threat to national and global security. Numerous reports on the linkages between climate change and security mention the potential instability resulting from massive movements of people displaced by climate-related impacts. In 2008, an official communication to the European Council on the issue noted that, "Europe must expect substantially increased migratory pressure." Yet rooting environmentally induced migration in a security agenda, and framing it in a deterministic perspective, is deeply at odds with the empirical realities of the climate change–migration nexus.[8]

The Impacts of Climate Change on Migration

The linkages between environmental changes and migration are extremely complex, and the relationship is far from direct or causal. Many uncertainties exist about the nature and strength of these linkages, in part because of the relative lack of empirical (particularly quantitative) studies. It is generally acknowledged that three types of climate change impacts can generate significant migration flows: sea-level rise, changes in precipitation patterns and associated water stress, and the increased intensity of natural hazards.[9]

Sea-level rise. The world's oceans are projected to rise by as much as one meter by the end of this century, although regional variation is expected. Coastal areas and river deltas rank among the most densely populated regions on Earth. They are home to many major cities—from Shanghai and Jakarta to London and New York—and will be at direct risk of flooding if adaptation measures, such as dikes and coastal restoration, are not implemented. Small-island nations are particularly vulnerable to even the slightest rise in sea level, which could inundate and eventually submerge buildings, roads, and other human structures.[10]

If no substantial adaptation measures are undertaken rapidly, people living in low-lying regions could be forced to relocate permanently, possibly abroad in the case of small-island developing countries. The time frame of these migrations, however, is very important: sea-level rise is a slow, incremental change, which allows populations to prepare and plan their relocation, possibly over several generations. In Kiribati, for example, the government has implemented a program called "Migration with Dignity," which aims to provide citizens with the necessary skills to migrate abroad by choice, before they are forced to do so.

Changes in precipitation patterns and associated water stress. Changes in precipitation and in the availability of water induce a different type of migration. Because water stress often mingles with other drivers of migration, such as poverty or land tenure issues, it is more difficult to assess the weight of environmental factors compared to other variables. Empirical research suggests that migration patterns might be more diversified, with people migrating both temporarily and permanently, typically from rural to urban areas. In sub-Saharan African countries, such as Niger, Benin, or Senegal, a member of a household often will migrate to the city to gain additional income and to sustain the household's livelihood during periods of drought, land degradation, and water stress. Remittances—the sending of money back home—are part of a household's strategy to cope with disruption in weather patterns.

Migration also can be part of a social routine to deal with environmental stress, as is often the case with livestock farming populations—but it can become a permanent relocation if the crisis becomes more severe, as has occurred in Kenya and South Sudan. In the latter case, pastoralist migration has been a trigger for conflict with sedentary populations. Likewise, severe droughts can induce brutal, dramatic displacements as people migrate in search of food and assistance. Yet there is also empirical evidence that the rate of migration can *decrease* in cases of extreme drought, because affected households are so diminished that they cannot afford to migrate. Migration patterns therefore depend greatly on the socioeconomic context, the assistance available, and the availability of migration options.[11]

Petterik Wiggers/Hollandse Hoogte

A child in a United Nations refugee camp in Juba, South Sudan.

Increased intensity of natural hazards. Extreme weather events, such as hurricanes and tornadoes, are expected to increase in intensity because of climate change—and they frequently trigger massive displacements of people. (See Box 9–1.) These displacements usually are confined within the borders of the affected country, but there have been cases of cross-border migration, especially when asylum possibilities were provided abroad. After Hurricane Mitch struck Central America in 1998, for example, many Hondurans and Nicaraguans were provided with temporary asylum in the United States.[12]

It was long thought that natural disasters did not lead to permanent migration, but only to temporary displacements, as the affected residents were assumed to return home once the disaster ended and reconstruction began. Hurricane Katrina, which devastated the U.S. Gulf Coast in 2005, however, showed that this was not always the case, as roughly a third of the population of New Orleans never returned to the city. But migration can also be a key tool for reconstruction in the aftermath of a disaster: remittances, for example, typically increase after a disaster, and can provide significant assistance to households seeking to rebuild their livelihoods.[13]

In general, migration patterns associated with environmental changes tend to be diverse and highly context-specific, making it difficult to outline common traits pertaining to "environmental migration." Yet various cross-country studies and reviews make it possible to draw general characteristics. The first is that most migrants move within the borders of their own country, and often for very short distances. This is because people usually have little interest in moving far away, as this would disrupt their economic and social networks and potentially deprive them of state-led assistance. Moreover, migration is a very costly endeavor, and many households simply do not have the resources to undertake an international migration.[14]

Another key characteristic is the intermingling of environmental factors with other drivers of migration, typically socioeconomic drivers such as poverty or job opportunities. Environmental factors cannot be set apart from their socioeconomic context, which makes it difficult to isolate a specific category of "environmental migration," with the exception of certain

Box 9–1. Natural Disasters and Human Displacement: Recent Trends

Natural disasters are displacing large numbers of people, although the numbers vary greatly from year to year. Population growth—the rise in overall human numbers as well as in the number of people exposed to hazards—has led to an increase in the scale of displacement over time. As the Internal Displacement Monitoring Centre observes, "improvements in disaster preparedness and response measures . . . mean that more people now survive disasters—but many of the survivors are displaced."

Demographics, vulnerability, and disaster risk reduction are key determinants of displacement, but environmental degradation and long-term climate change are becoming increasingly important. Although it remains difficult to pin a particular disaster on climate change, scientists have observed changes in both the magnitude and the frequency of extreme weather events in recent decades.

Weather-related hazards, such as floods, storms, and extreme temperatures, accounted for the vast majority (85 percent) of displacements due to rapid-onset natural disasters during 2008–13. Weather-related disasters displaced some 140 million people over this period, or an average of 23 million people per year. Floods (57 percent of the total) and storms (27 percent) were by far the most important contributors. (Earthquakes and volcanoes accounted for 15 percent of displacements in 2008–13, or slightly more than 24 million people.) (See Figure 9–1.)

But the annual figures fluctuate wildly along with variability in natural conditions. The number of people forced to flee in the face of weather hazards declined from 20.7 million in 2008 to 15.2 million in 2009, surged to 38.3 million in 2010, dropped to 13.8 million in 2011, more than doubled to 31.6 million in 2012, and then declined again to 20.6 million in 2013.

No comparable data exist on displacements caused by slow-onset disasters such as drought—whether they result from natural variability or are worsened by human-induced climate change. In extreme cases, lower or more-variable agricultural yields could compel people to move. Such migration may be seasonal, as people seek to supplement less-predictable

Figure 9–1. Population Displaced by Natural Disasters, by Disaster Type, 2008–2013

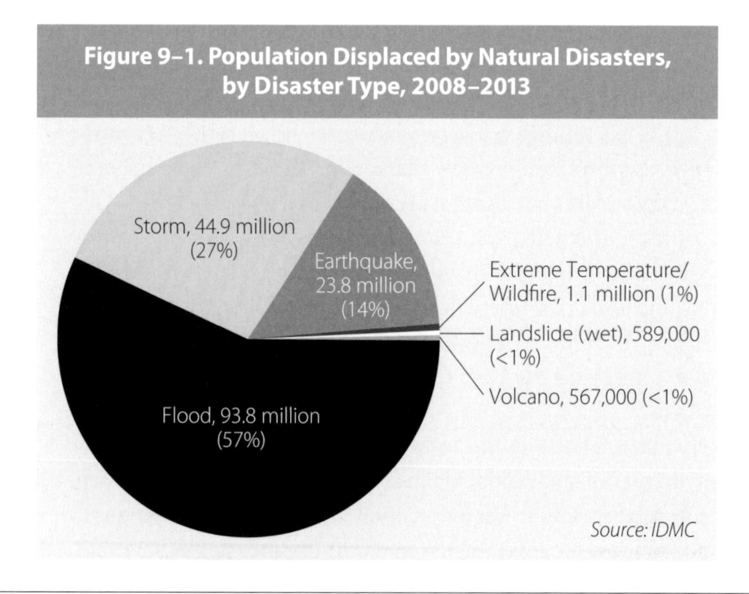

Source: IDMC

Box 9–1. continued

farm incomes with work elsewhere. But any such impacts are far less acute, and thus harder to capture in statistics, than impacts imposed by sudden-onset disasters. This is also true for another impact of climate change—sea-level rise—which may be gradual enough to permit counter-measures (such as dike construction) to obviate moving.

The number of people displaced due to climate impacts is expected to rise as extreme weather events become more frequent and intense, and as droughts, desertification, sea-level rise, and glacial melt become more prominent. Yet it seems impossible to make any reliable projections about how many people may be uprooted in the coming years and decades. There are simply too many unknowns: outcomes will depend on the precise nature of climate impacts, the time and place that disasters may strike, and the ways that disaster risks and impacts may be lessened by preparedness and adaptation.

Fast-onset impacts such as floods and storms affect people in different ways than more gradual (although perhaps longer-lasting) processes such as drought. The intensity and frequency of disasters may have different ramifications as well. And the impacts of one-time disasters may differ from the effects of successive catastrophes, such as the two typhoons and an earthquake that struck the Philippines within a four-month span during 2013, displacing nearly 6 million people. Overall, population movements in response to disasters can vary widely in their duration, characteristics, and destination. (See Figure 9–2.)

—Michael Renner

Source: See endnote 11.

Figure 9–2. Variations in Disasters and Population Movements

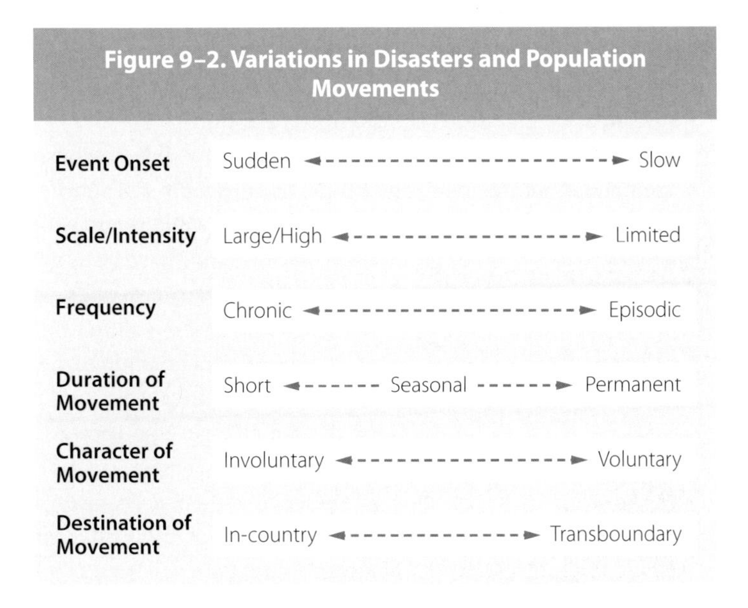

forced displacements associated with a brutal environmental disruption, such as a typhoon or flash flood. But even in these clear-cut cases, socioeconomic characteristics play an important role in determining the patterns of displacement. One cannot, for example, understand the patterns of displacement and return after Hurricane Katrina without taking into account race and poverty as determining factors.

The propensity to move is also highly dependent on age, gender, and wealth. Younger men tend to be more mobile than other categories, and the most vulnerable populations often find themselves unable to migrate. The poorest, in particular, frequently lack the resources that would allow them to afford the costs of transport, housing, and sometimes smuggling. When they do move, they typically travel shorter distances than wealthier populations, in some cases simply relocating from one hazardous zone to another, as has been documented in Bangladesh. These barriers to movement are primarily economic: depending on the locations involved, the cost of migrating from one country to another can amount to several years of a migrant's income.[15]

Administrative and informational barriers to movement also exist. In both industrialized and developing countries, migration policies have become increasingly stringent over time. Even when migrants move within their own country, they must overcome numerous administrative barriers, such as the possible loss of social benefits and protection. Many migrants lack information about possible destination areas and employment possibilities, and they often need to rely on migrants' networks to secure a livelihood. Land tenure is a critical issue as well: research shows that landowners, who often are reluctant to abandon their land, are less mobile than those who rent their land. Land tenure also is a frequent problem in the destination area, as migration can lead to competition for land.[16]

In a nutshell, migration is one of many possible responses to environmental disruption. (See Figure 9–3.) Some people will choose to migrate in order to adapt, while others will be forced to move because they have been unable to adapt. Some will adapt successfully in their home locales, while others will not be able to adapt at all, meaning that their lives, health, and livelihoods will be directly exposed to the impacts of climate change. The choice between these different options depends in part on the nature of the environmental changes, but also—and possibly more significantly—on the policy responses that are developed.

Figure 9–3. Climate Change Adaptation and Migration

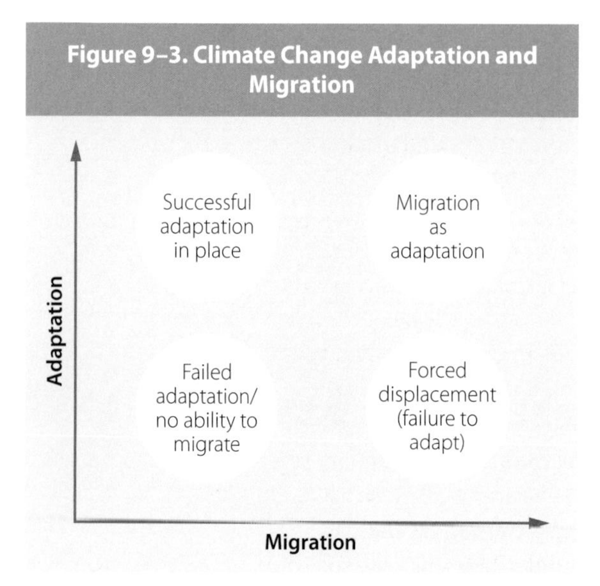

As planetary warming comes closer to topping the limit of a 2 degree Celsius (°C) increase in global average temperature, migration is likely to become a less viable adaptation option in the face of environmental changes, as people will likely have fewer adaptation opportunities from which to choose. In the event of a temperature increase above 4°C, both the number of people *forced* to move and the number of trapped people (*unable* to move) are likely to increase.[17]

At the moment, environmental changes are leading to both voluntary migration and forced displacement. Yet these categories are not discrete: most migration decisions include some elements of constraints, and very few movements are either completely voluntary or completely forced. In recent years, the line between voluntary and forced migration has become increasingly blurred, and forced and voluntary movements are better described as the two ends of a continuum than as clear-cut categories. In a warming world, where a particular population stands on this continuum will depend not only on climate impacts, but also—and possibly more importantly—on the way policies address climate-induced migration.[18]

Policy Directions for Migration as Adaptation

For the most part, climate-induced migration continues to be perceived as a failure of both migration and adaptation policies, and as a humanitarian catastrophe to be avoided at all cost. As a result, policy debates have focused mainly on protection and assistance mechanisms that could address this supposedly new type of migration. Yet empirical research shows that migration is not the only possible response that populations can adopt in the face of environmental disruption. Moreover, when they migrate, many people choose to do so willingly over other possible adaptation strategies: migration is employed as a powerful mechanism to diversify incomes, alleviate environmental pressures at home, send remittances, or simply put people and their families out of harm.

Yet the potential benefits of migration for adaptation should not overshadow the numerous situations of forced displacements, where people have no choice but to move because of environmental disruptions, such as persistent drought or submergence of their land. As climate change becomes more severe, with a global average temperature rise that could approach 4°C, migration is less likely to be available as an adaptation strategy. Consequently, more populations will find themselves either forced to migrate and relocate, or forced to stay where they are, because of a lack of resources and migration options.[19]

The paramount goal of policy responses should be to enable people's right to choose which adaptation strategy is best suited for their needs. This implies that people should be entitled with both the right to stay and the right to choose. Yet unabated climate change is likely to result in an increase not only in the number of forced migrants, but also in the number of "forced stayers."

Current adaptation policies tend to focus on the right to stay, with most projects targeted at the areas of origin affected by climate impacts. As such, migration is addressed mainly in a humanitarian or security agenda. Extending the migration options of populations, however, would require a broader development agenda. The right to choose one's adaptation strategy can be enabled only if people are provided with different migration options.

Two policy avenues should be considered in this regard. First, the most vulnerable populations should be provided with migration opportunities, including options that seek to address their lack of access to the resources, information, and networks that would allow them to relocate. If these populations are forced to stay where they are, they might find themselves directly exposed to climate dangers. Providing them with migration opportunities will require lifting numerous barriers to migration, including financial, informational, and administrative.

Second, adaptation policies should be directed toward the destination areas. These destinations often are urban areas in developing countries, whose possibilities to accommodate additional migrants may be limited. Significant adaptation efforts—related to infrastructure, land tenure, access to the job market and financial networks, etc.—will be needed within the host communities to ensure smooth integration of migrants.

Climate change induces both voluntary migration and forced displacement. While the latter can be seen as the symptom of overwhelmed adaptive capacities, the former can be regarded as a genuine adaptation strategy. But as climate impacts become more severe, it is likely that the migration options of vulnerable populations will be reduced significantly. A key challenge for adaptation policies will be to keep these options open, and to enable populations to choose their own adaptation strategies. Whether climate-induced migration is and will be an adaptation failure or an adaptation strategy depends not only on climate impacts, but—most importantly—on the policy choices that are made today.

Conclusion

CHAPTER 10

Childhood's End

Tom Prugh

When things stop working in an organizational system—a firm, a nonprofit, or a political entity—people have two choices in addressing the failure: to leave or to protest. Both can be powerful. As the author of this thesis, economist Albert Hirschman, pointed out in his 1970 book *Exit, Voice, and Loyalty*, it is possible for even an entire country (e.g., the United States or Liberia) to be created by people who leave behind unhappy circumstances and start something new elsewhere. Likewise, examples of the success of protest ("voice" and related action) in achieving major changes are plentiful. Consider the French Revolution and the multitude of regime changes and *coups d'état* that dot human history—not to mention the many times that a regime has been replaced by popular demand of voters.[1]

There is a point of scale, however, at which the choice of exit or voice shrinks simply to voice. We have reached that point, because the human sustainability dilemma now encompasses the entire biosocioeconomic system of Earth. Notwithstanding the avid fantasies of traveling to other planets to occupy them or to plunder their resources—as portrayed in such popular films as *Interstellar* and *Avatar*—there is nowhere else for us to go. Never mind the monumental technical obstacles to such travel; as biologist E. O. Wilson has pointed out, we are so intimately co-evolved with this utterly unique planetary ecosystem that no other, anywhere, would safely suit us.[2]

So it is a good sign that we humans increasingly are aware of and uneasy about the effects of our overuse and abuse of Earth, its habitats, its multitude of other creatures, its climate, and all the ecosystem services that it provides us. The fact is that we have gotten ourselves in a fine mess and we cannot emigrate our way out. Our only choice is how to come to grips with it. This drama can be seen, in effect, as a rite of passage for our species. To the extent that we succeed, we will have enshrined stewardship of Earth as our highest guiding principal and entered a phase of maturity something like adulthood.

It may be useful, in considering ways to address sustainability issues,

Tom Prugh is codirector of *State of the World 2015*.

to think in terms of tactical and strategic solutions. By "tactical," we mean specific policy actions or approaches aimed at particular aspects of the sustainability crisis. "Strategic" here means broader, overarching principles and ways of framing our relationship to the earth that could enhance the odds of achieving and maintaining a sustainable planetary civilization. Deployment of the tactical fixes cannot succeed in any lasting way unless it is informed, guided, and coordinated by a resolute adherence to the strategic principles (see below).

Tactical Solutions

The most prominent aspect of the mess we are in, of course, is climate change. The most recent assessment report from the Intergovernmental Panel on Climate Change (IPCC)—its fifth since 1990—only chisels in stone what has been evident for some time:

- The human influence on climate is "clear," and recent human-caused greenhouse gas emissions are "the highest in history."
- "Warming of the climate system is unequivocal, and since the 1950s, many of the observed changes are unprecedented over decades to millennia."
- The effects of those greenhouse gas emissions and other human drivers are "extremely likely" to have been the dominant cause of that warming.[3]

The warming documented by the IPCC drives many trends that are unwelcome at best and disastrous at worst: rising sea levels, increased species extinctions, greater weather extremes (droughts, floods, heat waves) and the resulting effects on people (hunger, famine, destruction of coastal communities), pest and disease vector migrations, social and political instability and conflict—and the triggering of positive feedback loops (carbon and methane releases from thawing tundra, for example) that spur further warming and threaten to tip the climate system into a state of uncontrollable derangement.

This book explores several related and worrying trends or issues, not yet fully on the public radar, that deepen the dimensions of our sustainability dilemma: the intersection of declining energy and rising debt; the growth-dependence of the global economy; stranded assets; agricultural resource loss; increasing ocean morbidity; the sociopolitical implications of a warming Arctic; the emergence of diseases from animals; and the challenge of climate-induced migration. What might be called "tactical measures" are already available to confront all of these problems.

The immediate means of addressing climate change, for example, are by now as well known as the problem. Among the most useful responses would be a global agreement acknowledging the gravity of climate change and pledging action to curb the (mostly) fossil fuel emissions that cause it.

Eric Kort, Jet Propulsion Laboratory

Cracks in Arctic sea ice north of Alaska. Research flights have detected higher methane levels above open water than over sea ice. This could represent a noticeable new global source of methane.

In December 2014, the nations of the world gathered in Lima, Peru (at the twentieth session of the Conference of the Parties to the United Nations Framework Convention on Climate Change) to lay the groundwork for such an agreement, which, with any luck, will be concluded in December 2015 in Paris. This meeting was widely said to have been energized by an agreement reached a few weeks before between the United States and China, the two biggest carbon polluters, to set targets for limiting their emissions, although many observers were disappointed in the prospects for a strong outcome in Paris. One noted that the language of "commitments" was weakened to "contributions," that the need for solid language on adaptation went unaddressed, and that the terms of the "contributions" and their assessment were left largely unspecified.[4]

Several additional negotiation sessions are scheduled to take place prior to the December meeting in Paris, so perhaps some of these weaknesses can still be rectified. Assuming that there is an actual treaty in 2015, signatories should be able to choose from a wide array of available technical, social, and economic options for developing their detailed emission reduction plans and meeting their pledge targets. Commitment of funding to help developing nations cope with climate change is uncertain, but it is to be hoped that they will have access to technical and financial assistance from the industrial nations, which historically are responsible for most of the warming to date.

Likewise, each of the hidden threats to sustainability discussed in this book has already begun to engage capable minds in its solution, and options are plentiful. For example:

- **The energy, growth, and credit nexus** (Chapter 2). Addressing the risks associated with declining energy supplies and quality, and rising costs and debt, could entail banking reform, carbon and/or consumption taxes, replacing GDP as the top measure of economic well-being, and a broad spectrum of physical and psychological preparations and education for the decline and end of growth-oriented economies.
- **Uneconomic growth** (Chapter 3). It may be helpful to remind ourselves that the pursuit of economic growth as a policy objective is only a few decades old and is not an inherent property of economies. Although economic stagnation is not desirable either, an economy can be dynamic

without growing. Policies should aim to reduce the material throughput that demands resources and energy, and to improve equitable distribution, which will have the effect of reining in growth. Options include incentives and policies for converting productivity gains into leisure rather than increased consumption, incentives to restore soils and habitats, curbs on financial speculation, and incentives for investment in public goods (infrastructure, community facilities, natural amenities).

- **Stranded assets** (Chapter 4). Environmental trends, market prices for commodities, technologies, government regulations, social perceptions, and other factors can affect the current and future value of assets held by firms, including capital investments (such as power plants) and inventories of lands and resources (such as oil or minerals). Policies and management practices to mitigate these risks include recognizing and understanding the process of value destruction and creation, avoiding technology and infrastructure lock-in, aiming for the smooth offsetting of value lost by value creation, and the exposure and proper pricing of environment-related risks.
- **Agricultural resource loss** (Chapter 5). The shrinking availability of land and water for food production at a time of rising population and demand for food can be offset in several ways: reducing food waste (perhaps one-third of global production is wasted each year at various stages); increasing water productivity by careful tracking of water efficiency and selection of crops according to the regional abundance of water; using conservation easements and other tools to avoid loss of farmland to development; and reducing the production of meat and biofuels.
- **Ocean morbidity** (Chapter 6). The health and productivity of the oceans are at risk first from climate change (warming and acidification), second from overfishing, and finally from synergies between the two. The tactical means of addressing climate change are discussed above. Overfishing can often be addressed by means of conservation structures such as firmly enforced marine protected areas and abolition of fishing equipment and techniques that destroy ocean-bottom habitats and result in bycatch.
- **Managing Arctic challenges** (Chapter 7). The profound changes occurring in the Arctic, largely because of climate change, are caused in the main by people who do not live there, and non-Arctic interests—both development-oriented and activist—have tended to pursue their own agendas in the region with little regard for local interests and wishes. The long-term sustainability of the region requires acknowledgment of the ability and right of local and indigenous peoples to make their own decisions about development and conservation.

- **Emerging diseases from animals** (Chapter 8). The transmission of diseases from animals to humans is among the most alarming but least recognized public health trends of recent years. Ebola, SARS, hantavirus, monkeypox, Lyme disease, Nipah virus, and other diseases originating in wildlife account for most of the emerging infectious diseases in humans. The complexity of the interactions among many factors means that a multidisciplinary approach is required, particularly involving collaboration among ecologists, clinicians, public health scientists, and governmental and intergovernmental agencies to model risks, predict emergence of new diseases, and monitor disease incidence in animals.

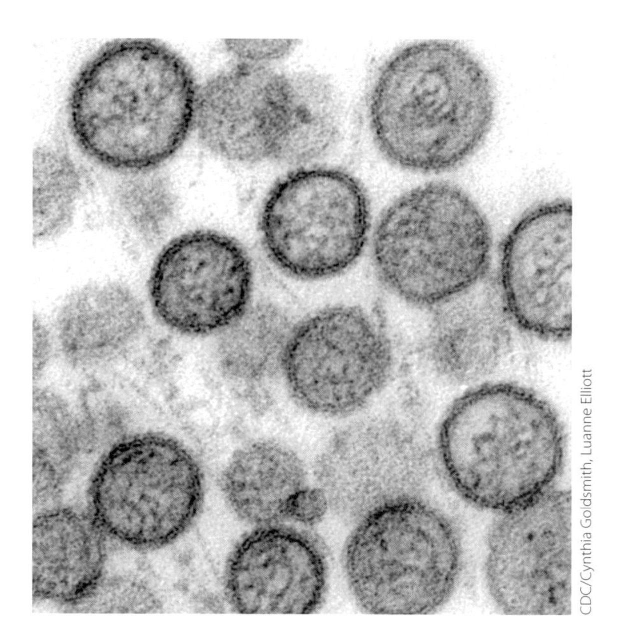

CDC/Cynthia Goldsmith, Luanne Elliott

Transmission electron micrograph of hantavirus.

- **Coping with climate migrants** (Chapter 9). Although predictions of tens or hundreds of millions of "climate refugees" may still be premature, it is clear that climate change will induce some degree of both voluntary migration and forced displacement. The patterns and circumstances of migration are more complex than generally acknowledged, however, and far from always signifying disaster, such movements can be viable adaptation options. Policies to address these issues can reduce barriers to migration for the most vulnerable populations as well as enhance the ability of destination areas to accommodate migrants and enable their smooth integration.

Strategic Solutions

Tactical options may be incomplete, inappropriate, or simply ignored unless there is a strategic vision impelling and coordinating their use. A vision of "sustainability," in turn, needs to be supported by strategies, or "meta-solutions," that provide structure in framing and approaching the problems of sustainability in an integrated way. We discuss three such meta-solutions below.

Systems thinking. One key strategy is systems thinking. For example, because the scale of human influence on the biosphere is now global, we must come to think routinely of ourselves and our economies as nested subsystems that are embedded in the global ecosystem, unviable apart from it. This is an example of a "pre-analytic vision" that re-frames a problem and throws light on fresh ways of approaching it. The pre-analytic vision at issue here is from ecological economics, which holds that economic growth is essentially the process of converting more and more of nature to things that we want. Put this way, it is clear that converting all of nature to things we want—stuff—is impossible, as that would destroy things that we need in order to live at all.[5]

The very language that we use (even in this book) reveals how deeply entrenched is our traditional model of nature as big-box store. Phrases such as "marshaling the earth's resources" betray a mindset that views the planet as a warehouse of raw stuff available for the taking. But viewing the earth as a complex and dynamic system of subsystems and energy flows reveals that what we really do through economic growth is not just take a tree off the shelf to make lumber for a house, or fish from the sea to make dinner, but subtract living parts of a system. Although the Earth system is complex and deeply resilient, we have been doing this for thousands of years, with increasingly large numbers of humans busily pursuing their own projects that subtract parts from the system. The result is that we now tremble on the brink of serious compromise of the biosphere.

SeppVei

A forest harvester at work in a Finnish pine forest.

Shifting to a systems-thinking mode clearly leads to a much broader definition of "environmentalism." It is not enough to focus on simply cleaning up polluted rivers, or shutting down coal-fired power plants so that their mercury emissions stop poisoning downwind communities. Sociopolitical developments in recent years, especially but not exclusively in the United States, have helped to lock in certain destructive forces that ultimately threaten the prospects for genuine sustainability. Gus Speth, the former dean of the Yale School of Forestry and Environmental Studies, who was instrumental in founding the World Resources Institute and the Natural Resources Defense Council, puts it this way:

> We've got to ask afresh, "What is an environmental issue?" The conventional answer is air and water pollution, climate change, and so on. But what if our answer is: "Whatever determines environmental outcomes." Once we think about it this way, then, surely, the creeping plutocracy and corporatocracy we face—the ascendency of money power and corporate power over people power—these are environmental issues. And

> more: The chartering and empowering of artificial persons to do virtually anything in the name of profit and growth—that is the very nature of today's corporation; the fetish of GDP growth as the ultimate public good and the main aim of government; our runaway consumerism; our vast social insecurity with half the families living paycheck to paycheck. These are among the underlying drivers of environmental outcomes. To succeed, . . . environmentalists are going to have to address these issues.[6]

Thinking this way is challenging and can lead to both unexpected insights and resistance. For example, nearly 30 years ago, nutrition experts Joan Gussow and Katherine Clancy published an article suggesting that "educated consumers need to make food choices that not only enhance their own health but also contribute to the protection of our natural resources"—in other words, that what people choose to eat (especially how much meat they want in their diets, a demand that is soaring worldwide) can profoundly shape agricultural practices and thereby affect, for good or ill, the viability of the biosphere. Within the environmental community, this wisdom is now commonplace. However, it was not until late 2014 that the U.S. Dietary Guidelines Advisory Committee took up the matter and invited Clancy to testify. The U.S. Congress, with the protests of meat industry trade groups ringing in their ears, quickly passed a "congressional directive" instructing the panel to ignore the diet/sustainability connection in revising the guidelines.[7]

One systems framework that may be particularly relevant for analyzing human social, political, and economic interactions with the biosphere is the theory of panarchy. Developed by Canadian ecologist C. S. Holling from his careful observations of forest ecology, panarchy theory describes complex systems in terms of their cycles of development. Considerably oversimplified, it proposes that such systems—including the socioecologic system that comprises humans and our interactions with the biosphere—unfold in four adaptive phases: growth, collapse, regeneration, and growth again. In the growth phase, a system becomes progressively more complex, integrated, and efficient, but also less resilient, i.e., more brittle and less able to absorb shocks or disturbances and still bounce back. Eventually, a shock arrives—a fire in a forest, perhaps, or a globally significant financial crisis—which the system in question is unable to handle. The consequent collapse can be partial and mild or deep and violent (or something in between), but any collapse frees up resources that can be recombined in novel ways during the next growth phase. While inflicting hardship, collapse thus also presents opportunities for renewal.[8]

The point here is that with globalization have come bigger institutions, tighter economic and financial integration across national boundaries, longer supply lines, just-in-time manufacturing delivery systems, greater social complexity, and myriad other developments that suggest rising efficiency but declining resilience—and the increasing prospect of sharp and painful contractions and/or upheavals as the system is buffeted by the inevitable shocks. These outcomes can be mitigated and softened, but only if citizens and policy makers are aware of the process and prepared to make adjustments ahead of the crisis point.

Stewardship. A second framing device or meta-solution is the notion that humanity needs to cultivate an attitude of stewardship toward the earth rather than one of domination, control, and exploitation. This is hardly a new viewpoint, but the need for it has become increasingly urgent as the consequences of unbridled growth have become more evident. Stewardship follows naturally from the worldview of Earth as a rich and fertile but bounded ecosystem rather than an infinite warehouse, and also acknowledges that science-fiction tales of human emigration to the stars are dubious "solutions" at best.

Stewardship is both practical and virtuous. It acknowledges that this is the only place we live, and it implies the need for ongoing care of the planetary ecosystem so as to maintain its capacity to support and nurture all of humanity indefinitely. As ecological economist Herman Daly has written, it thereby also implies "an extension of brotherhood" to future generations as well as to the multitudes of other creatures that are our "coevolutionaries" and with which we share the planet.[9]

Implicit in the idea and practice of stewardship is maturity. By and large, it is to adults, or nascent adults, that the role of caretaker falls. Some children and many adolescents may show remarkable signs of maturity at an early age, but it is not truly expected of them until they are a few years older. Some, of course, never seem to reach maturity at all. There is an important place for them too—the energy, exuberance, and brashness often typical of adolescents (and reflected in the Silicon Valley mantra "move fast and break things") will never be useless; we will likely need all the do-it-now spirit and creativity we can muster to solve the problems now facing us and lying ahead. But somebody needs to keep a steady hand on the tiller and a view toward the horizon, and those qualities somehow need to be made indelible in our species. Simply wishing for that virtue is not likely to achieve it, but what if something like it arose from an institutionally structured process?

Robust Citizenship

If there are any candidates for such a process, then surely a better, stronger, deeper, more responsive, and more widespread democracy—our third meta-solution—is one, perhaps even the best one. Worldwatch has argued before (see *State of the World 2014: Governing for Sustainability*) that a shift in our political systems toward grassroots empowerment may be a potent way to take up the challenges of sustainability:

Steven Lyons, Credo Action

A San Francisco demonstration against the Keystone XL pipeline.

> A democracy of distributed leadership (as opposed to one that begins and ends with the ballot box) seems to be the natural home—if such a new idea as sustainability can be said to have one—for sustainability efforts. . . . Where democracy is already in place, citizens and civil society organizations need to take advantage of their existing freedoms to organize, protest, deliberate, offer input to governments, and demand action. Where democracy is mainly for show or simply absent, safer tactics are required. The goal is the same: to create the irresistible force needed to elicit a positive response.[10]

Grassroots democracy, as expressed in mass action, has a long and inspiring history of significant successes, from the dogged struggle to abolish slavery begun in the sixteenth century, to the U.S. civil rights movement, the anti-Apartheid movement in South Africa, and the continuing campaign to secure suffrage and equal rights for women around the globe. It can also claim success on the environmental front: if it were not for Earth Day, for instance, there likely would not be a U.S. Clean Air Act or any of the other pillars of modern environmental regulation. More recently, the Keystone XL pipeline, intended to carry landlocked tar sands bitumen from Alberta, Canada, to refineries on the U.S. Gulf Coast, has suffered from a concerted popular effort to stall and kill it. An October 2014 report suggests that the anti-Keystone forces have helped force the cancellation of several tar sands projects, cut capital expenditures by the development firms, and reduce tar sands producer revenues by nearly $31 billion.[11]

Movements can be difficult to sustain, but it is also possible to build grassroots governance structures that encourage deeper engagement by ordinary people in processes that lead to better-informed and more politically viable solutions to difficult problems. By such structures, we do not necessarily mean the typical republican or parliamentary systems now prevalent in most nominal democracies. Those systems undoubtedly are better than autocracy, but they have major flaws and mock the promise of deeper democratic practices. According to political theorist James Fishkin, such systems have empowered people but

> . . . under conditions in which the people have little reason or effective incentive to think very much about the power we would ask them to exercise. . . . [T]he mass public in almost every polity lacks information or does not even pay much attention to political matters. And when the public is mostly uninformed, it is easily subject to manipulation by the mechanisms of one-sided persuasion developed for advertising.[12]

If we want more than that—something worthy of the label "democracy," something other than a "sound-bite democracy of manipulation and electoral advantage" (in Fishkin's damning phrase)—we must look to the practices of participatory and deliberative democracy. Fortunately, those are not fanciful theories in political science textbooks, but concrete and widely employed means of engaging great numbers of ordinary people in solving problems and making policy.

They might almost be called hidden solutions—with the potential to address the hidden threats discussed in this book as well as the rest of the sustainability problems that communities and nations face—because they rarely appear in the mainstream media and are largely unknown to most people, even those who live their entire lives in countries where democracy means little more than voting every few years. Many of the places where deliberative and other deep forms of democracy are practiced most vigorously lie in the global South, where democracies are younger and people are more interested in experimenting with its forms and institutions. For example, participatory budgeting (in which ordinary people deliberate and decide how to spend public monies) was essentially invented in Porto Alegre, Brazil. Examples of such sustained engagement can be found all over the world.[13]

Matt Leighninger of the Deliberative Democracy Consortium argues that "it is the lack of strong democracy that underlies so many of the . . . challenges we face. Failing schools, friction between citizens and police, urban sprawl, incivility and hyperpartisanship in politics, structural racism, conflicts over immigration, unworkable local budgets . . . they are all symptoms of the inability of public institutions to react to, and capitalize on, what citizens want and can do." Strong democracy can mobilize the talents and energies of large numbers of people whose interests are directly at stake in the issues they confront.[14]

Strong democracy may also be the best way of attacking one of the most neglected, even untouchable, social justice dimensions of sustainability:

inequality in wealth and income. Building cultures of local and regional democracy will generate higher-quality decisions and policy about most problems that communities perpetually face, but inequality is of particular concern because it lies at the root of many social ills that affect rich and poor alike. By a host of measures—physical and mental health, life expectancies, educational performance, rates of violence and imprisonment, social mobility, and others—well-being for all members of a society is improved when it strives to limit the kind of vast inequalities of wealth, income, and power that so often have characterized human cultures and threaten to overtake even those with well-established middle classes.[15]

United Workers

Discussion of growing inequality at a workshop of the Economic Democracy Conference in Baltimore.

In fact, strong local and regional democracies may be the only competent antidote to such inequalities, if human history is any guide. This particular issue may take on a special urgency as we approach the end of the fossil fuel era because of the effect of abundant energy on social and political structures. (See Box 10–1.)[16]

A thriving culture of local democracies—perhaps aggregated into regional and even national assemblies in a way that retains a deliberative and participatory character—would be a good thing even if the future turns out to be less energy-poor than seems likely. But especially if the future is one of relative energy poverty, communities wishing to control both their environmental and political fates would do well to establish grassroots democratic structures now, so that they are deeply rooted enough to withstand the winds of change during the turbulent transition to a low-energy future.

There is no question that scholars and scientists who study the human economy, the earth, and the interactions between them are drawing profoundly troubling conclusions. The trends discussed in this book, which are unfolding before our eyes in real time, are nothing to be sanguine about. They reflect clear and present dangers, not worries that can safely be deferred until some vague future date. To address them, we need to learn stewardship, not escape (whether that escape is to another planet or into mindless consumerism).

No one can predict the future. But if the human species has any concern for its fate on Earth, then there is much to ponder in the sober views of those who believe that the thrill ride of explosive economic and population

Box 10–1. Fossil Energy and the Global Middle Class

Most major societies since the development of agriculture, some 10,000 years ago, have been organized around production for surplus. In contrast to the small and "flat" societies of hunter-gatherer bands, agricultural societies grew in size and produced steep social hierarchies with wealthy rulers, warriors, priests, and artisans at the top and masses of poor laborers at the bottom. With rare exceptions, that arrangement seemed like most peoples' lot in life, and statistically it was, bolstered by elaborate justificatory philosophies and values systems such as the "divine right" of kings.

It was the expropriated mass labor—energy—of the people at the bottom (hundreds or thousands for every member of the elite) that enabled the elite to enjoy leisure and luxury. And so things stood for thousands of years, until something new happened: fossil fuels. As discussed in Chapter 2, when we learned how to extract fossil fuels and developed the technology to leverage their capacity to do work, we in effect captured millions of "slaves" and put them to work for us. This flood of cheap "labor" helped wreak a profound change in the distribution of wealth, at least in industrial societies, in the form of large middle classes. Over the last couple of centuries, hundreds of millions of people have had the good fortune to be born in a time when they could live in the kind of comfort and wealth unknown to all but the elites of the pre-fossil era.

Because political power follows wealth, with those middle classes came considerable power and eventually mass democracy. But if power follows wealth, and wealth hinges on available energy, what happens to power when the energy from fossil fuels declines?

The answer seems likely to be "back to the future." In energy-poor eras (i.e., most of our history), the pattern of tiny rich elites ruling impoverished masses seems almost ubiquitous in human societies. As Ronald Wright notes in *A Short History of Progress*:

> When the Spaniards reached the American mainland in the sixteenth century, the peoples of the western and eastern hemispheres had not met since their ancestors parted as Ice Age hunters running out of game. . . . Two cultural experiments, running in isolation for 15,000 years or more, at last came face to face. Amazingly, after all that time, each could recognize the other's institutions. When Cortés landed in Mexico he found roads, canals, cities, palaces, schools, law courts, markets, irrigation works, kings, priests, temples, peasants, artisans, armies, astronomers, merchants, sports, theatre, art, music, and books.

In a post-fossil era, with energy supplies sharply reduced, there is little reason to believe that allocation of the available wealth will not revert to its traditional "civilized" pattern: a few very rich, most very poor. Likewise, a shrunken or vanished middle class means a re-concentration of political power at the top. A post-fossil era would seem just as vulnerable to this plight as any society that existed prior to the Industrial Revolution—especially when we seem to be moving in that direction already.

Source: See endnote 16

growth, by which we are heedlessly turning the entire planet to our own purposes and in the process ravaging it, is coming to an end. The nature of that end is, with every passing day, less and less a matter of our choosing. It is time for *Homo sapiens sapiens* to live up to its somewhat presumptuous Latin name, and grow up.

Notes

Chapter 1. The Seeds of Modern Threats

1. People's Climate March, "Wrap-Up," http://peoplesclimate.org/wrap-up/; Hansen testimony from Philip Shabecoff, "Global Warming Has Begun, Expert Tells Senate," *New York Times*, June 24, 1988.

2.. "History of Climate Chance Science," *Wikipedia*, http://en.wikipedia.org/wiki/History_of_climate_change_science; World Meteorological Organization, "Declaration of the World Climate Conference" (Geneva: February 1979).

3. Hansen quote from Shabecoff, "Global Warming Has Begun, Expert Tells Senate."

4. Intergovernmental Panel on Climate Change, "Human Influence on Climate Clear, IPCC Report Says," press release (Geneva: September 27, 2013).

5. Robert Engelman, "Beyond Sustainababble," in Worldwatch Institute, *State of the World 2013: Is Sustainability Still Possible?* (Washington, DC: Island Press, 2013).

6. Upton Sinclair, *I, Candidate for Governor: And How I Got Licked* (1935) (Oakland, CA: University of California Press, 1994 reprint).

7. Naomi Klein, *This Changes Everything: Capitalism vs. The Climate* (New York: Simon & Schuster, 2014).

8. Organisation for Economic Co-operation and Development, *Towards Green Growth* (Paris: 2011), 9.

9. Walter V. Reid et al., *Millennium Ecosystem Assessment: Ecosystems and Human Well-Being*, Synthesis Report (Washington, DC: Island Press, 2005); fish stocks from United Nations Environment Programme (UNEP), *Towards a Green Economy: Pathways to Sustainable Development and Poverty Eradication* (Nairobi: 2011), 20; Robert J. Diaz and Rutger Rosenberg, "Spreading Dead Zones and Consequences for Marine Ecosystems," *Science* 321, no. 5891 (August 15, 2008): 926–29; "What Causes Ocean 'Dead Zones'?" *Scientific American*, September 25, 2012; pollinators from Gary Gardner and Thomas Prugh, "Seeding the Sustainable Economy," in Worldwatch Institute, *State of the World 2008* (New York: W. W. Norton & Company, 2008), 3; World Health Organization, "7 Million Premature Deaths Annually Linked to Air Pollution," press release (Geneva: March 25, 2014).

10. J. R. McNeill, *Something New Under the Sun: An Environmental History of the Twentieth-Century World* (New York: W. W. Norton & Company, 2000), 11.

11. Ibid., 13.

12. Mid-1700s from Gary T. Gardner, *Inspiring Progress* (New York: W. W. Norton & Company, 2006); Science Intelligence and InfoPros, "How Many Science Journals?" January 23, 2012, http://scienceintelligence.wordpress.com/2012/01/23/how-many-science-journals/; Rose Eveleth, "Academics Write Papers Arguing Over How Many People Read (And Cite) Their Papers," *Smithsonian*, March 25, 2014; Richard Van Noorden, "Global Scientific Output Doubles Every Nine Years," *Nature Newsblog*, May 7, 2014, http://blogs.nature.com/news/2014/05/global-scientific-output-doubles-every-nine-years.html.

13. Gardner, *Inspiring Progress*.

14. See John M. Gowdy, "Governance, Sustainability, and Evolution," in Worldwatch Institute, *State of the World 2014: Governing for Sustainability* (Washington, DC: Island Press, 2014), 31–40; McNeill, *Something New Under the Sun*, xxiv.

15. McNeill, *Something New Under the Sun*, 15.

16. Ibid., 14, 31; data for 2000 and 2013 from BP, *BP Statistical Review of World Energy 2014* (London: 2014).

17. Gardner, *Inspiring Progress*.

18. Compiled from U.S. Geological Survey, *Mineral Commodity Summaries* (Reston, VA: various editions), and from International Iron and Steel Institute, *A Handbook of World Steel Statistics* (Brussels: 1978), 1.

19. Quotation, 1970–2010 trend, and China from UNEP, *Global Chemicals Outlook – Towards Sound Management of Chemicals* (Nairobi: 2013), xv, 11; number of compounds synthesized and in use from McNeill, *Something New Under the Sun*, 28; doubling from European Chemical Industry Council (Cefic), *The European Chemical Industry: Facts & Figures 2013* (Brussels: January 2014).

20. UNEP, *Global Chemicals Outlook*, xv, 10.

21. McNeill, *Something New Under the Sun*, 25–26; 2000 data from Matt Styslinger, "Fertilizer Consumption Declines Sharply," *Vital Signs Online* (Worldwatch Institute), 21 October 2010; 2013 estimate from Patrick Heffer and Michel Prud'homme, *Fertilizer Outlook 2013–2017*, prepared for the 81st International Fertilizer Industry Association Conference, Chicago, IL, May 20–22, 2013, 4.

22. Production in 1900 from "The Automobile Industry, 1900–1909," The History of Technology Website, Bryant University, Smithfield, RI, http://web.bryant.edu/~ehu/h364/materials/cars/cars%20_10.htm; Michael Renner, "Auto Production Sets New Record, Fleet Surpasses 1 Billion Mark," *Vital Signs Online* (Worldwatch Institute), June 4, 2014; fleet in 1900 from "History of Motor Car/Automobile Production 1900–2003," www.carhistory4u.com/the-last-100-years/car-production; 1910 from McNeill, *Something New Under the Sun*, 60; 1960 from Michael Renner, *Rethinking the Role of the Automobile*, Worldwatch Paper 84 (Washington, DC: June 1988); 2013 from Colin Couchman, IHS Automotive, London, e-mail to author, May 27, 2014.

23. Changes in emissions calculated from U.S. Environmental Protection Agency, "1970–2013 Average annual emissions, all criteria pollutants in MS Excel," February 2014, www.epa.gov/ttn/chief/trends/trends06/national_tier1_caps.xlsx; Table 1–1 adapted from McNeill, *Something New Under the Sun*, 54.

24. World Steel Association, *World Steel in Figures 2014* (Brussels: 2014), 7, 9; 2000 calculated from International Iron and Steel Institute, *Steel Statistical Yearbook 2002* (Brussels: 2002), 12.

25. Doubling of production from UNEP, *Keeping Track of Our Changing Environment. From Rio to Rio+20 (1992–2012)* (Nairobi: October 2011); total extraction trend from Sustainable Europe Research Institute, GLOBAL 2000, and Friends of the Earth Europe, *Overconsumption? Our Use of the World's Natural Resources* (Vienna: September 2009), 9–10; Table 1–2 from UNEP, *Keeping Track of Our Changing Environment*; passenger car statistics calculated from data provided by Couchman, IHS Automotive.

26. Table 1–3 adapted from Chris Bright, "Anticipating Environmental 'Surprise,'" in Worldwatch Institute, *State of the World 2000* (New York: W. W. Norton & Company, 2000), Tables 2–1 and 2–2.

27. Nick Breeze, "Global Warming – You Must Be Joking! How Melting Arctic Ice Is Driving Harsh Winters," *The Ecologist*, November 21, 2014.

28. McNeill, *Something New Under the Sun*, 62; Magda Lovei, *Phasing Out Lead from Gasoline. Worldwide Experiences and Policy Implications*, World Bank Technical Paper No. 397, Pollution Management Series (Washington, DC: 1998); 2011 achievements from Peter Lehner, "Global Phase-out of Lead in Gasoline Succeeds: Major Victory for Kids' Health," Switchboard blog (Natural Resources Defense Council), October 27, 2011.

29. "Smog," *Science Daily*, www.sciencedaily.com/articles/s/smog.htm; "Smog," *Wikipedia*, http://en.wikipedia.org/wiki/Smog.

30. Molly O. Sheehan, "CFC Use Declining," in Worldwatch Institute, *Vital Signs 2002* (New York: W. W. Norton & Company, 2002), 54–55; Alana Herro, "Ozone Layer Stabilizing But Not Recovered," in Worldwatch Institute, *Vital*

Signs 2007–2008 (New York: W. W. Norton & Company, 2007), 45–46; UNEP, "Ozone Layer on Track to Recovery: Success Story Should Encourage Action on Climate," press release (Nairobi: September 10, 2014).

31. Heinrich Böll Stiftung and Friends of the Earth Europe, *Meat Atlas: Facts and Figures About the Animals We Eat* (Berlin and Brussels: 2014), 26; "Another Strike Against GMOs – The Creation of Superbugs and Superweeds," GMO Inside blog, March 31, 2014; Tom Laskawy, "First Came Superweeds; Now Come the Superbugs!," *Grist*, March 25, 2010.

32. Thomas Homer-Dixon, *The Upside of Down: Catastrophe, Creativity, and the Renewal of Civilization* (Washington, DC: Island Press, 2006), 253. Also see Homer-Dixon's discussion in Chapter 9 of his book for an extended discussion of cycles of growth, collapse, regeneration, and renewed growth in the context of forest systems. Similar characteristics can be observed in human societies.

Chapter 2. Energy, Credit, and the End of Growth

1. U.S. Congressional Budget Office, *The Budget and Economic Outlook: 2014 to 2024* (Washington, DC: February 4, 2014); William H. Gross, "Investment Outlook: For Wonks Only" (Newport Beach, CA: PIMCO, September 2014).

2. U.S. Census Bureau, *Income and Poverty in the United States: 2013* (Washington, DC: September 2014); U.S. Energy Information Administration (EIA), "Petroleum & Other Liquids: Product Supplied," www.eia.gov/dnav/pet/pet_cons_psup_dc_nus_mbblpd_a.htm, updated November 26, 2014; Doug Short, "Vehicle Miles Traveled: A Structural Change in Our Behavior," October 20, 2014, www.advisorperspectives.com/dshort/updates/DOT-Miles-Driven.php; Jan Luiten van Zanden et al., eds., *How Was Life? Global Well-being Since 1820* (Paris: Organisation for Economic Co-operation and Development, 2014); Blake Ellis, "28% of Americans Have No Emergency Savings," *CNN Money*, June 25, 2012.

3. The Editors of Encyclopædia Britannica, "Trophic Pyramid," *Encyclopædia Brittanica*, updated February 3, 2013, www.britannica.com/EBchecked/topic/606499/trophic-pyramid.

4. Robert W. Howarth, "Coastal Nitrogen Pollution: A Review of Sources and Trends Globally and Regionally, *Harmful Algae* 8, no. 1 (December 2008): 14–20; Figure 2–1 from the following sources: energy consumption compiled by author from BP, *Statistical Review of World Energy* (London: 2011), from Vaclav Smil, *Energy Transitions: History, Requirements, Prospects* (Praeger: Santa Barbara, CA, 2010), from Suneeta D. Fernandes et al., "Global Biofuel Use 1850–2000," *Global Biogeochemical Cycles* 21, no. 2 (June 2007), and from IHS Energy; population data for 1800–1950 from United Nations Population Division, *The World at Six Billion* (New York: 1999), 5; population data for 1960–2012 from U.S. Census Bureau, International Programs, "World Population," www.census.gov/population/international/data/worldpop/table_population.php, viewed November 17, 2014.

5. Box 2–1 from ejolt, "Human Energy Use (Endosomatic/Exosomatic)," www.ejolt.org/2012/12/human-energy-use-endosomatic-exosomatic/, and from "General Laborer Salaries," Salary.com, http://www1.salary.com/General-Laborer-salary.html; Table 2–1 adapted from Institute for Integrated Economic Research (IIER), "Green Growth: An Oxymoron?" July 31, 2011, www.iier.ch/content/green-growth-oxymoron.

6. Figure 2–2 from the following sources: GDP from U.S. Department of Agriculture, Economic Research Service, "Real GDP (2010 dollars) Historical," International Macroeconomic Data Set, www.ers.usda.gov/data-products/international-macroeconomic-data-set.aspx, viewed November 2014; electricity from EIA, "Total Electricity Net Generation," International Energy Statistics, www.eia.gov/cfapps/ipdbproject/IEDIndex3.cfm?tid=2&pid=2&aid=12, viewed November 2014, road and marine transportation fuels from ExxonMobil, *The Outlook for Energy: A View to 2040* (Irving, TX: 2014 and previous years); primary energy from EIA, "Total Primary Energy Production, International Energy Statistics, www.eia.gov/cfapps/ipdbproject/IEDIndex3.cfm?tid=44&pid=44&aid=1, viewed November 2014. Note that electricity generation and primary energy production data are used as an approximation for consumption levels.

7. IIER, *Low Carbon and Economic Growth – Key Challenges* (Meilen, Switzerland: July 2011).

8. IIER, "Green Growth: An Oxymoron?" 18; Max Kingsley-Jones, "Emirates Begins Parting Out Its A340-500s," *Flight International*, September 23, 2013; EIA, "The Cement Industry Is the Most Energy Intensive of All Manufacturing Industries," *Today in Energy*, July 1, 2013, www.eia.gov/todayinenergy/detail.cfm?id=11911.

9. Megan C. Guilford et al., "A New Long Term Assessment of Energy Return on Investment (EROI) for U.S. Oil and Gas Discovery and Production," *Sustainability* 3, no. 10 (2011): 1866–87.

10. N. Beveridge et al., *Era of Cheap Oil Over As Secular Growth in Upstream Cost Inflation Underpins Triple Digit Oil Prices* (New York: Bernstein Energy, 2012); Goldman Sachs, *Higher Long-Term Prices Required for Troubled Industry* (New York: April 12, 2013); Russell Gold and Erin Ailworth, "Fracking Firms Get Tested by Oil's Price Drop," *Wall Street Journal*, October 9, 2014. Box 2–3 from the following sources: natural gas shortfall in Kevin Landfried, Jeffrey Steimer, and Barbara Weber, "Bridging the Gap," *LNG Industry*, Autumn 2005; Mike Ruppert, "Interview with Matthew Simmons," August 18, 2003, www.oilcrash.com/articles/blackout.htm; horizontal drilling from Dan O'Brien, "Longest Lateral: Consol Innovates Efficiencies," *Business Journal* (Youngstown, OH), November 16, 2013; U.S. oil and gas production from BP, *Statistical Review of World Energy 2014* (London: 2014), 8, 24; McKinsey Global Institute, *Game Changers: Five Opportunities for US Growth and Renewal* (New York: 2013); Thomas H. Darrah et al., "Noble Gases Identify the Mechanisms of Fugitive Gas Contamination in Drinking-water Wells Overlying the Marcellus and Barnett Shales," *Proceedings of the National Academy of Sciences* 111, no. 39 (September 30, 2014): 14076–81; Joe Eaton, "Oklahoma Grapples with Earthquake Spike—and Evidence of Industry's Role," *National Geographic News*, July 31, 2014.

11. Tyler Durden, "The Imploding Energy Sector Is Responsible for a Third of S&P 500 Capex," Zero Hedge, November 30, 2014, www.zerohedge.com/news; EIA, "Oil and Gas Industry Employment Growing Much Faster Than Total Private Sector Employment," *Today in Energy*, August 8, 2013, www.eia.gov/todayinenergy/detail.cfm?id=12451.

12. Tad W. Patzek, "Thermodynamics of the Corn-Ethanol Biofuel Cycle," *Critical Reviews in Plant Sciences* 23, no. 6 (2004): 519–67; Platts, *Special Report: New Crudes, New Markets* (New York: March 2013).

13. David J. Murphy and Charles A. S. Hall, "Energy Return on Investment, Peak Oil, and the End of Economic Growth," in Robert Costanza, Karin Limburg, and Ida Kubiszewski, eds., "Ecological Economics Reviews," *Annals of the New York Academy of Sciences* 1219 (February 2011): 52–72.

14. Michael McLeay, Amar Radia, and Ryland Thomas, "Money Creation in the Modern Economy," *Quarterly Bulletin* (Bank of England), 2014 Q1.

15. Josh Ryan-Collins et al., *Where Does Money Come From? A Guide to the UK Monetary and Banking System* (London: New Economics Foundation, September 29, 2011).

16. McLeay, Radia, and Thomas, "Money Creation in the Modern Economy."

17. Bank for International Settlements, "Debt Securities Statistics," www.bis.org/statistics/secstats.htm; EIA, *Monthly Energy Review* (Washington, DC: December 2014), 151. Box 2–4 from the following sources: Sam Ro, "Here Are the Breakeven Oil Prices for America's Shale Basins," *Business Insider*, October 22, 2014, www.businessinsider.com/shale-basin-breakeven-prices-2014-10; Bob Pisani, "Here's Why Oil Stocks are Tanking," *CNBC*, October 10, 2014, www.cnbc.com/id/102078976; APICORP Research, "OPEC in the Future: Will It Continue to Play a Pivotal Role?" *Economic Commentary* 9, no. 10 (October 2014), Table 11; Kristen Hays, "Exclusive: New U.S. Oil and Gas Well November Permits Tumble Nearly 40 Percent," *Reuters*, December 2, 2014.

18. Global Economic Intersection, "$15 Trillion Central Bank Balance Sheets (Fictitious Capital?)," January 31, 2012, http://econintersect.com/b2evolution/blog1.php/2012/01/31/15-trillion-central-bank-balance-sheets-fictitious-capital.

Chapter 3. The Trouble with Growth

1. Institute for Sustainable Development and International Relations, "An Innovative Society for the Twenty-first Century" conference, Paris, July 12–13, 2013; goals from Dan O'Neill, "A Post-Growth Economy in France?" Center for Advancement of a Steady-State Economy blog, July 2013, http://steadystate.org/a-post-growth-economy-in-france/.

2. Intergovernmental Panel on Climate Change, "Reports," www.ipcc.ch/publications_and_data/publications_and_data_reports.shtml, viewed October 28, 2014; Millennium Ecosystem Assessment, *Living Beyond Our*

Means: Natural Assets and Human Well-being: Statement from the Board (March 2005); Johan Rockstrom et al., "Planetary Boundaries: Exploring the Safe Operating Space for Humanity," *Ecology and Society* 14, no. 2 (September 2009); WWF, *Living Planet Report 2014* (Gland, Switzerland: 2014).

3. Heinz W. Arndt, *The Rise and Fall of Economic Growth: A Study in Contemporary Thought* (Melbourne: Longman Cheshire, 1978), 30.

4. John Maynard Keynes, *The General Theory of Employment Interest and Money* (New York: Harcourt-Brace, 1936).

5. Organisation for Economic Co-operation and Development (OECD), *Convention on the Organisation for Economic Co-operation and Development* (Paris: OECD, 1960).

6. Peter A.Victor, ed., *The Costs of Economic Growth.* (Cheltenham, U.K.: Edward Elgar, 2013).

7. Herman E. Daly and John B. Cobb, *For the Common Good: Redirecting the Economy Toward Community, the Environment, and a Sustainable Future* (Boston: Beacon Press, 1994).

8. Daly and Cobb, *For the Common Good*; Ida Kubiszewski et al., "Beyond GDP: Measuring and Achieving Global Genuine Progress," *Ecological Economics* 93 (September 2013): 57–93.

9. Herman E. Daly, *Steady-State Economics: The Economics of Biophysical Equilibrium and Moral Growth* (San Francisco: W. H. Freeman, 1977).

10. OECD, *Economic Policy Reforms: Going for Growth* (Paris: 2014); OECD, *Towards Green Growth. A Summary for Policy Makers* (Paris: 2011).

11. Tim Jackson, *Material Concerns: Pollution, Profit and the Quality of Life* (Oxon, U.K.: Routledge, 1996); Julian M. Allwood and Jonathan M. Cullen, *Sustainable Materials – With Both Eyes Open* (Cambridge, U.K.: UIT Cambridge, 2012).

12. Global Commission on the Economy and Climate, *Better Growth Better Climate. The New Climate Economy Report. The Synthesis Report* (Washington, DC: 2014); Gunter Pauli, *The Blue Economy. !0 Years, 100 Innovations, 100 Million Jobs* (Taos, NM: Paradigm Publications, 2010); Ellen MacArthur Foundation, *Towards the Circular Economy, Vols. 1–3* (Cowes, Isle of Wight, U.K: 2012–14).

13. Factor 10 Institute, www.factor10-institute.org. Before Factor 10 there was Factor 4; see Ernst von Weizsäcker, Amory B. Lovins, and L. Hunter Lovins, *Factor Four: Doubling Wealth, Halving Resource Use* (London: Earthscan, 1997). After Factor 10, there came Factor 5; see Ernst von Weizsäcker et al., *Factor 5: Transforming the Global Economy through 80% Improvements in Resource Productivity* (Oxon, U.K.: Routledge, 2009).

14. Vaclav Smil, *Making the Modern World: Materials and Dematerialization* (West Sussex, U.K.: Wiley, 2013); Smil uses dematerialization rather than decoupling, but for consistency decoupling is used throughout the chapter except where Smil is quoted directly; Tim Jackson, *Prosperity Without Growth: Economics for a Finite Planet* (London: Earthscan, 2009).

15. William S. Jevons, *The Coal Question: An Inquiry Concerning the Progress of the Nation, and the Probable Exhaustion of Our Coal-mines* (New York: A. M. Kelley, 1865); Mona Chitnis et al., "Who Rebounds Most? Estimating Direct and Indirect Rebound Effects for Different UK Socioeconomic Groups," *Ecological Economics* 106 (October 2014): 12–32; Mona Chitnis et al., "Turning Lights into Flights: Estimating Direct and Indirect Rebound Effects for UK Households," *Energy Policy* 55 (April 2013): 234–50.

16. Smil, *Making the Modern World*, 180. Smil's primary focus is on materials other than fossil fuels, although when he does discuss energy and decarbonization he says that despite the ongoing process of relative decarbonization, "there is no imminent prospects for any major reductions in absolute emissions of CO_2 . . ." (ibid, 154).

17. Angela Druckman and Tim Jackson, "The Carbon Footprint of UK Households 1990-2004: A Socio-economically Disaggregated, Quasi-multiregional Input-Output Model," *Ecological Economics* 68, no. 7 (May 2009): 2066–77.

18. Thomas O. Wiedmann et al., "The Material Footprint of Nations," *Proceedings of the National Academy of Sciences*, August 2013, 1–6; M. Lenzen et al., "Mapping the Structure of the World Economy," *Environmental Science*

& Technology 46, no. 15 (2012): 8374–81; M. Lenzen et al., "Building Eora: A Global Multi-Region Input-Output Database at High Country and Sector Resolution," *Economic Systems Research* 25, no. 1 (2013): 20–49.

19. Wiedmann et al., "The Material Footprint of Nations," 4.

20. John Stuart Mill, *Principles of Political Economy: With Some of Their Applications to Social Philosophy* (London: Penguin Books, 1970), 113–14.

21. Ibid., 114, 116.

22. Peter A. Victor, *Managing Without Growth. Slower by Design, Not Disaster* (Cheltenham, U.K.: Edward Elgar, 2008); Tim Jackson, *Prosperity Without Growth.*

23. Tim Jackson and Peter A. Victor, "Productivity and Work in the 'Green Economy': Some Theoretical Reflections and Empirical Tests," *Environmental Innovation and Societal Transitions* 1, no. 1 (June 2011): 101–08; Tim Jackson and Peter A. Victor, *Green Economy at Community Scale* (Toronto, ON: The Metcalf Foundation, 2013); Tim Jackson and Peter A. Victor, *Does Low Growth Increase Inequality?* (Guildford, U.K.: University of Surrey, 2014).

24. Victor, *Managing Without Growth.* Figure 3–2 based on idem, 182.

25. Giorgos Kallis, Christian Kerschner, and Joan Martinez-Alier, "The Economics of Degrowth," *Ecological Economics* 84 (December 2012): 172–80; Peter A. Victor, "Growth, Degrowth and Climate Change: A Scenario Analysis," *Ecological Economics* 84 (December 2012): 206–12; Giacomo D'Alisa, Federico Demaria, and Giorgos Kallis, eds., *Degrowth: A Vocabulary for a New Era* (London: Routledge, 2014).

26. Thomas Piketty, *Capital in the Twenty-First Century* (Cambridge, MA: The Belknap Press of Harvard University Press, 2014).

Chapter 4. Avoiding Stranded Assets

1. Ben Caldecott, James Tilbury, and Yuge Ma, *Stranded Down Under? Environment-Related Factors Changing China's Demand for Coal and What This Means for Australian Coal Assets* (Oxford, U.K.: Smith School of Enterprise and the Environment, Oxford University, 2013).

2. Ibid.; The White House, "U.S.-China Joint Announcement on Climate Change," press release (Washington, DC: November 11, 2014).

3. Smith School of Enterprise and the Environment, Oxford University, "Stranded Assets Programme: Introduction," www.smithschool.ox.ac.uk/research/stranded-assets.

4. Ibid.; Table 4–1 based on Ben L. Caldecott, Nicholas Howarth, and Patrick McSharry, *Stranded Assets in Agriculture: Protecting Value from Environment-Related Risks* (Oxford, U.K.: Smith School of Enterprise and the Environment, Oxford University, 2013).

5. Carbon Tracker, *Unburnable Carbon: Are the World's Financial Markets Carrying a Carbon Bubble?* (London: 2011); Bill McKibben, "Global Warming's Terrifying New Math," *Rolling Stone*, July 19, 2012.

6. Carbon Tracker, *Unburnable Carbon.*

7. Atif Ansar, Ben L. Caldecott, and James Tilbury, *Stranded Assets and the Fossil Fuel Divestment Campaign: What Does Divestment Mean for the Valuation of Fossil Fuel Assets?* (Oxford, U.K.: Smith School of Enterprise and the Environment, Oxford University, 2013); HSBC Global Research, *Coal and Carbon. Stranded Assets: Assessing the Risk* (London: 2012); Elad Jelasko, *Carbon Constraints Cast a Shadow Over the Future of the Coal Industry* (London: Standard & Poor's Ratings Services, July 2014); Justin Gillis, "U.N. Panel Warns of Dire Effects From Lack of Action Over Global Warming," *New York Times*, November 3, 2014.

8. Lesley Fleischman et al., "Ripe for Retirement: An Economic Analysis of the U.S. Coal Fleet," *The Electricity Journal* 26, no. 10 (2013): 51–63; P. Knight et al., *Forecasting Coal Unit Competitiveness: Coal Retirement Assessment Using Synapse's Coal Asset Valuation Tool (CAVT)* (Cambridge, MA: Synapse Energy Economics, 2013).

9. U.S. Environmental Protection Agency, "EPA Proposes First Guidelines to Cut Carbon Pollution from Existing

Power Plants," press release (Washington, DC: June 2, 2014); U.S. Energy Information Administration, *Monthly Energy Review* (Washington, DC: July 2014).

10. Bloomberg New Energy Finance, *2030 Market Outlook* (London: 2013).

11. Fleischman et al., "Ripe for Retirement: An Economic Analysis of the U.S. Coal Fleet."

12. Barney Jopson and Ed Crooks, "Obama Proposes Biggest Ever US Push for Carbon Cuts," *Financial Times*, June 2, 2014; Atkins quote from Mark Chediak and Jim Polson, "Obama Climate Proposal Will Shift Industry Foundations," *Bloomberg News*, June 2, 2014.

13. Knight et al., *Forecasting Coal Unit Competitiveness*; Thomas Tindall and Fabien Roques, *Keeping Europe's Lights on: Design and Impact of Capacity Mechanisms* (London: IHS Cambridge Energy Research Associates, Inc., 2013).

14. Louis-Mathieu Perrin, *Benchmarking European Power and Utility Asset Impairments* (Paris: EY, 2013), 5.

15. Mark C. Lewis, *Stranded Assets, Fossilised Revenues: USD28trn of Fossil-Fuel Revenues at Risk in a 450-ppm World* (Paris: Kepler Cheuvreux and ESG Sustainability Research, April 24, 2014).

16. Carbon Tracker, *Unburnable Carbon*, 32.

17. Nafeez Ahmed, "The Inevitable Demise of the Fossil Fuel Empire," *The Guardian* (U.K.), June 10, 2014; Mark Fulton and Reid Capalino, "Trillion-dollar Question: Is Big Oil Over-investing in High-cost Projects?" RenewEcon omy.com.au, May 21, 2014.

18. Goldman Sachs, *380 Projects to Change the World: From Resource Constraint to Infrastructure Constraint* (New York: 2013), 126.

19. Box 4–2 from Millennium Ecosystem Assessment, *Ecosystems and Human Well-being: Synthesis* (Washington, DC: Island Press, 2005), and from Trucost Plc, *Natural Capital at Risk: The Top 100 Externalities of Business*, prepared for the TEEB for Business Coalition (London: April 15, 2013).

20. Caldecott, Tilbury, and Ma, *Stranded Down Under?*

21. Brian J. Henderson, Neal D. Pearson, and Li Wang, *New Evidence on the Financialization of Commodity Markets*, updated November 18, 2014, http://ssrn.com/abstract=1990828; Ing-Haw Cheng and Wei Xiong, *The Financialization of Commodity Markets*, NBER Working Paper No. 19642 (Cambridge, MA: National Bureau of Economic Research, November 2013); Troy Sternberg, "Chinese Drought, Wheat, and the Egyptian Uprising: How a Localized Hazard Became Globalized," in Caitlin E. Werrell and Francesco Femia, eds., *The Arab Spring and Climate Change: A Climate and Security Correlations Series* (Washington, DC: Center for American Progress, 2013).

22. Sternberg, 7.

23. George Welton, "The Impact of Russia's 2010 Grain Export Ban," *Oxfam Policy and Practice: Agriculture, Food and Land* 11, no. 5 (2011): 76–107; J. A. Lampietti et al., "A Strategic Framework for Improving Food Security in Arab Countries," *Food Security* 3, no. 1 (2011): 7–22; Index Mundi, "Wheat Imports by Country in 1000 MT," 2013, www.indexmundi.com/agriculture/?commodity=wheat&graph=imports; André Croppenstedt, Maurice Saade, and Gamal M. Siam, "Food Security and Wheat Policy in Egypt," *Roles of Agriculture Project Policy Brief* (Rome: United Nations Food and Agriculture Organization, October 2006).

24. Table 4–2 from Trucost Plc, *Natural Capital at Risk.*

25. World Bank, *Global Economic Prospects: Commodities at the Crossroads* (Washington, DC: 2009); Savills Research – Rural, *International Farmland Focus 2012* (London: 2012), 4. Although converting to U.S. dollars per hectare can have an effect on annual growth rates in terms of domestic currency, it does allow potential investors a good starting point for comparable analysis.

26. Intergovernmental Panel on Climate Change (IPCC), *Climate Change 2014: Impacts, Adaptation and Vulnerability. Contribution of Working Group II to the Fifth Assessment Report of the Intergovernmental Panel on Climate Change* (Cambridge, U.K. and New York: Cambridge University Press, 2014).

27. Zbigniew W. Kundzewicz et al., "Chapter 3: Freshwater Resources and Their Management," in M. L. Parry et al.,

eds., *Climate Change 2007: Impacts, Adaptation and Vulnerability. Contribution of Working Group II to the Fourth Assessment Report of the Intergovernmental Panel on Climate Change* (Cambridge, U.K. and New York: Cambridge University Press, 2007), 187; Eleanor Burke, Simon Brown, and Nikolaos Christidis, "Modelling the Recent Evolution of Global Drought and Projections for the 21st Century with the Hadley Centre Climate Model," *Journal of Hydrometeorology* 7 (2006): 1113–25; T. P. Barnett, J. C. Adam, and D. P. Lettenmaier, "Potential Impacts of a Warming Climate on Water Availability in Snow Dominated Regions," *Nature* 438 (November 17, 2005): 303–09; Bryson Bates et al., eds., *Climate Change and Water*, IPCC Technical Paper VI (Geneva: IPCC, 2008).

28. Box 4–3 from Ben Caldecott, "The Solution to Coal Plants? Pay Their Owners to Close Them," ChinaDialogue .net, September 19, 2014.

29. Gordon L. Clark, Andreas Feiner, and Michael Viehs, *From the Stockholder to the Stakeholder: How Sustainability Can Drive Financial Outperformance* (Oxford, U.K.: Smith School of Enterprise and the Environment, Oxford University, October 20, 2014).

Chapter 5. Mounting Losses of Agricultural Resources

1. Richard Howitt et al., *Economic Analysis of the 2014 Drought for California Agriculture* (Davis, CA: Center for Watershed Sciences, University of California, Davis, July 2014); 5 percent is a Worldwatch calculation based on total irrigated farmland in California, from the following sources: about 10 million acres in 2000 from Susan S. Hutson et al., *Estimated Use of Water in the United States in 2000* (Reston, VA: U.S. Geological Survey, updated February 2005), Table 7; about 9 million acres from American Farmland Trust, "California Drought Increases Need to Conserve Farmland," www.farmland.org/programs/states/ca/CA-Drought-NeedtoConserveFarmland.asp, viewed December 10, 2014.

2. Susanne Moser et al., *Our Changing Climate 2012: Vulnerability & Adaptation to the Increasing Risks from Climate Change in California* (Sacramento, CA: California Climate Change Center, July 2012); Dan Cayan et al., *Scenarios of Climate Change in California: An Overview* (Sacramento, CA: California Climate Change Center, February 2006).

3. Hectares and San Franciscos based on acreage conversion data from California Department of Conservation, *California Farmland Conversion Report 2008–2010* (Sacramento, CA: April 2014), and from U.S. Census Bureau, "State and County QuickFacts," http://quickfacts.census.gov/qfd/states/06/06075.html, viewed October 8, 2014.

4. Nikos Alexandratos and Jelle Bruinsma, *World Agriculture Towards 2030/2050: The 2012 Revision* (Rome: United Nations Food and Agriculture Organization (FAO), June 2012), 61, 95; United Nations Environment Programme (UNEP), "UNEP's Emerging Issues" (Nairobi: 2011), www.unep.org/pdf/RIO20/UNEP-%20Emerging -Issues.pdf.

5. FAO, International Fund for Agricultural Development, and World Food Programme, *The State of Food Insecurity in the World 2014. Strengthening the Enabling Environment for Food Security and Nutrition* (Rome, FAO, 2014).

6. Biofuel shares from FAO, "The State of Food and Agriculture," conference document (Rome: June 2013).

7. Alexandratos and Bruinsma, *World Agriculture Towards 2030/2050*; FAO, "Global Capture Production" and "Global Aquaculture Production," Fishery Statistical Collections, electronic databases, www.fao.org/fishery/statis tics/en, viewed November 2014.

8. Alexandratos and Bruinsma, *World Agriculture Towards 2030/2050*; FAO, "Coping with Water Scarcity in the Near East and North Africa," fact sheet prepared for Regional Conference for the Near East, Rome, February 24–28, 2014, www.fao.org/docrep/019/as215e/as215e.pdf.

9. Malin Falkenmark, "Growing Water Scarcity in Agriculture: Future Challenge to Global Water Security," *Philosophical Transactions of the Royal Society A* 371, no. 2002 (November 13, 2013).

10. Arjen Y. Hoekstra, "Global Monthly Water Scarcity: Blue Water Footprints versus Blue Water Availability," *PLOS ONE* 7, no. 2 (2012); "Egypt to 'Escalate' Ethiopian Dam Dispute," *Al Jazeera*, April 21, 2014.

11. Stephen Foster and Tushaar Shah, *Groundwater Resources and Irrigated Agriculture: Making a Beneficial Relation More Sustainable* (Stockholm: Global Water Partnership, 2012); UNEP, "A Glass Half Empty: Regions at Risk

Due to Groundwater Depletion," January 2012, www.unep.org/pdf/UNEP-GEAS_JAN_2012.pdf; Falkenmark, "Growing Water Scarcity in Agriculture"; Katalyn A. Voss et al., "Groundwater Depletion in the Middle East from GRACE with Implications for Transboundary Water Management in the Tigris-Euphrates-Western Iran Region," *Water Resources Research* 49, no. 2 (February 2013): 904–14; Cynthia Barnett, "Groundwater Wake-up," *Ensia* (University of Minnesota), August 19, 2013.

12. Table 5–1 based on World Bank, "Renewable Internal Freshwater Resources per Capita (cubic meters)," electronic database, http://data.worldbank.org/indicator/ER.H2O.INTR.PC; Jacob Schewe et al., "Multimodel Assessment of Water Scarcity Under Climate Change," *Proceedings of the National Academy of Sciences* 111, no. 9 (December 16, 2013): 3245–50.

13. United Nations Department of Economic and Social Affairs, International Decade for Action "Water for Life" 2005–2015, "Water Scarcity," www.un.org/waterforlifedecade/scarcity.shtml.

14. Water scarcity from World Bank, "Renewable Internal Freshwater Resources per Capita (cubic meters)," electronic database, http://data.worldbank.org/indicator/ER.H2O.INTR.PC; import dependence a Worldwatch calculation based on data from U.S. Department of Agriculture (USDA), Foreign Agricultural Service (FAS), "Production, Supply, and Distribution," electronic database, https://apps.fas.usda.gov/psdonline/psdQuery.aspx, viewed September 15, 2014.

15. Arjen Y. Hoekstra and Mesfin M. Mekonnen, "The Water Footprint of Humanity," *Proceedings of the National Academy of Sciences* 109, no. 9 (February 28, 2012).

16. Ibid.; Arjen Y. Hoekstra, "Water Security of Nations: How International Trade Affects National Water Scarcity and Dependency," *Threats to Global Water Security, NATO Science for Peace and Security Series C: Environmental Security* (2009): 27–36.

17. FAO, *The State of the World's Land and Water Resources for Food and Agriculture*; Alexandratos and Bruinsma, *World Agriculture Towards 2030/2050*, 105.

18. L. R. Oldeman, R. T. A. Hakkeling, and W. G. Sombroek, *World Map of the Status of Human-Induced Soil Degradation: An Explanatory Note* (Wageningen: International Soil Reference and Information Centre and Nairobi: UNEP, October 1990); Z. G. Bai et al., "Proxy Global Assessment of Land Degradation," *Soil Use and Management* 24, no. 3 (September 2008): 223–34; FAO, *The State of the World's Land and Water Resources for Food and Agriculture.*

19. Fred Pearce, "Splash and Grab: The Global Scramble for Water," *New Scientist*, March 2, 2013, 28–29; Japan area of 377,915 square kilometers from U.S. Central Intelligence Agency, "Country Comparison: Area," in *The World Factbook*, https://www.cia.gov/library/publications/the-world-factbook/rankorder/2147rank.html, viewed August 13, 2014; shares based on Land Matrix, "Intention of Investment," http://landmatrix.org/en/get-the-idea/dynamics-overview/, viewed September 24, 2014; Table 5–2 from Land Matrix, "Web of Transnational Deals," http://landmatrix.org/en/get-the-idea/web-transnational-deals/, viewed November 13, 2014.

20. Lorenzo Cotulo, *Land Deals in Africa: What Is in the Contracts?* (London: International Institute for Environment and Development, 2011); Brian Bienkowski and Environmental Health News, "Corporations Grabbing Land and Water Overseas," *Scientific American*, February 12, 2013; Table 5–3 from Land Matrix, "Web of Transnational Deals."

21. Bienkowski and Environmental Health News, "Corporations Grabbing Land and Water Overseas"; Ward Anseeuw et al., *Land Rights and the Rush for Land: Findings of the Global Commercial Pressures on Land Research Project* (Rome: International Land Coalition, 2012).

22. John R. Porter and Liyong Xie, "Chapter 7. Food Security and Food Production Systems," in Intergovernmental Panel on Climate Change (IPCC), *Climate Change 2014: Impacts, Adaptation, and Vulnerability, Working Group II Contribution to the Fifth Assessment Report of the Intergovernmental Panel on Climate Change* (Cambridge, U.K.: Cambridge University Press. 2014); Gerald C. Nelson et al., *Climate Change: Impact on Agriculture and Costs of Adaptation* (Washington, DC: International Food Policy Research Institute, October 2009); Cynthia Rosenzweig et al., "Assessing Agricultural Risks of Climate Change in the 21st Century in a Global Gridded Crop Model Intercomparison," *Proceedings of the National Academy of Sciences* 111, no. 9 (March 4, 2014): 3268–73.

23. Christopher B. Field et al., "Summary for Policymakers," in IPCC, *Climate Change 2014: Impacts, Adaptation, and Vulnerability*; Porter and Xie, "Chapter 7. Food Security and Food Production Systems."

24. Joshua Elliott et al., "Constraints and Potentials of Future Irrigation Water Availability on Agricultural Production Under Climate Change," *Proceedings of the National Academy of Sciences* 111, no. 9 (March 4, 2014): 3239–44

25. Porter and Xie, "Chapter 7. Food Security and Food Production Systems."

26. Table 5–4 based on USDA, FAS, "Production, Supply, and Distribution."

27. Figure 5–1 and Central America, Middle East and North Africa, and Japan are Worldwatch calculations based on data from USDA, FAS, "Production, Supply, and Distribution"; other regions from Stacey Rosen, USDA, Economic Research Service, personal communication with author, December 9, 2014.

28. Hoekstra and Mekonnen, "The Water Footprint of Humanity."

29. Jenny Gustavsson et al., *Global Food Losses and Food Waste: Extent, Causes, and Prevention* (Rome: FAO, 2011); UNEP, "Food Waste Facts" (Nairobi: 2013), www.unep.org/wed/2013/quickfacts/.

30. CGIAR, "Postharvest Loss Reduction – A Significant Focus of CGIAR Research," November 20, 2013, www.cgiar.org/consortium-news/postharvest-loss-reduction-a-significant-focus-of-cgiar-research/; Gustavsson et al., "Global Food Losses and Food Waste"; Aramark Higher Education, *The Business and Cultural Case for Trayless Dining* (Philadelphia, PA: July 2008).

31. UNEP, "Food Waste Facts."

32. Mesfin M. Mekonnen and Arjen Y. Hoekstra, "Water Footprint Benchmarks for Crop Production: A First Global Assessment," *Ecological Indicators* 46 (November 2014): 214–23.

33. Table 5–5 from Ibid.

34. USDA, "Production, Supply, and Distribution"; Table 5–6 from Arjen Y. Hoekstra, "The Water Footprint of Animal Products," in J. D'Silva and J. Webster, eds., *The Meat Crisis: Developing More Sustainable Production and Consumption* (London: Earthscan, 2010); J. Liu and H. H. G. Savenije, "Food Consumption Patterns and Their Effect on Water Requirement in China," *Hydrology and Earth System Sciences* 12 (2008): 887–98.

35. Arjen Y. Hoekstra, "Water for Animal Products: A Blind Spot in Water Policy," *Environmental Research Letters* 9 (2014); Arjen Y. Hoekstra, "The Hidden Water Resource Use Behind Meat and Dairy," *Animal Frontiers* 2, no. 2 (April 2012): 3–8.

36. Estimates of 30 and 40 percent from USDA, *USDA Agricultural Projections to 2022* (Washington, DC: February 2013); Rosamond Naylor, "Biofuels, Rural Development, and the Changing Nature of Agricultural Demand," paper presented at the Symposium Series on Global Food Policy and Food Security, Stanford University, Palo Alto, CA, April 11, 2012; FAO, "The State of Food and Agriculture."

37. Marianela Fader, "Spatial Decoupling of Agricultural Production and Consumption: Quantifying Dependences of Countries on Food Imports Due to Domestic Land and Water Constraints," *Environmental Research Letters* 8, no. 1 (2013). See also Appendix A of the article, which cites additional studies that offer other perspectives, and Miina Porkka et al., "From Food Insufficiency Towards Trade Dependency: A Historical Analysis of Global Food Availability," *PLOS ONE* 8, no. 12 (December 18, 2013); FAO, *The Right to Food: Past Commitment, Current Obligation, Further Action for the Future: A Ten-Year Retrospective on the Right to Food Guidelines* (Rome: 2014).

Chapter 6. The Oceans: Resilience at Risk

1. Herman Melville, *Moby-Dick* (Berkeley and Los Angeles: University of California Press, 1979).

2. National Oceanic and Atmospheric Administration (NOAA), U.S. Department of Commerce, "Brief History of the Groundfish Industry of New England," www.nefsc.noaa.gov/history/stories/groundfish/grndfsh1.html, viewed November 4, 2014.

3. NOAA National Ocean Service, "How Much of the Ocean Have We Explored?" http://oceanservice.noaa.gov/facts/exploration.html, viewed October 12, 2014.

4. Estimate of 3 billion from United Nations Food and Agriculture Organization (FAO), *Global Aquaculture Advancement Partnership (GAAP) Programme*, prepared for the Seventh Session of the COFI Sub-Committee on Aquaculture, St. Petersburg, Russia, October 7–11, 2013; LIFDCs from "The Role of Seafood in Global Food Security," advance and unedited reporting material on the topic of focus of the fifteenth meeting of the United Nations Open-ended Informal Consultative Process on Oceans and the Law of the Sea, March 14, 2014, www.un.org/depts/los/consultative_process/documents/adv_uned_mat.pdf; Table 6–1 from Gertjan de Graaf and Luca Garibaldi, *The Value of African Fisheries* (Rome: FAO, 2014); Joint Ocean Commission Initiative, *America's Ocean Future: Ensuring Healthy Oceans to Support a Vibrant Economy* (Washington, DC: June 2011).

5. Figure 6–1 from FAO, *The State of World Fisheries and Aquaculture* (Rome: 2014).

6. California Environmental Associates, *Charting a Course to Sustainable Fisheries* (San Francisco: July 2012); FAO, *The State of World Fisheries and Aquaculture.*

7. Boris Worm et al., "Impacts of Biodiversity Loss on Ocean Ecosystem Services," *Science* 314, no. 787 (November 3, 2006): 787–90.

8. National Marine Fisheries Service, Northeast Fisheries Science Center, *Gulf of Maine Atlantic Cod: 2014 Assessment Update* (Woods Hole, MA: August 22, 2014).

9. Dennis M. King and Jon G. Sutinen, "Rational Noncompliance and the Liquidation of Northeast Groundfish Resources," *Marine Policy* 34 (2010): 7–10; Pew Charitable Trusts, *Risky Decisions: How Denial and Delay Brought Disaster to New England's Historic Fishing Grounds* (Philadelphia, PA: October 2014).

10. Katie Auth, *Fishing for Common Ground: Broadening the Definition of "Rights-based" Fisheries Management in Iceland's Westfjords* (Akureyri, Iceland: University of Akureyri, 2012).

11. Boris Worm et al., "Rebuilding Global Fisheries," *Science* 325, no. 5940 (July 31, 2009): 578–85; California Environmental Associates, *Charting a Course to Sustainable Fisheries.*

12. E. Griffin et al., *Predators as Prey: Why Healthy Oceans Need Sharks* (Washington, DC: Oceana, 2008); Ransom A. Myers et al., "Cascading Effects of the Loss of Apex Predatory Sharks from a Coastal Ocean," *Science* 315, no. 5820 (March 30, 2007): 1846–50.

13. Bettina Wassener, "China Says No More Shark Fin Soup at State Banquets," *New York Times*, July 3, 2012; FAO, *The State of World Fisheries and Aquaculture*, 122.

14. FAO, *The State of World Fisheries and Aquaculture.*

15. California Environmental Associates, *Charting a Course to Sustainable Fisheries.*

16. Global Carbon Project, *Global Carbon Budget 2014* (Canberra, Australia: September 21, 2014).

17. Xianyao Chen and Ka-Kit Tung, "Varying Planetary Heat Sink Led to Global-warming Slowdown and Acceleration," *Science* 345, no. 6199 (August 22, 2014): 897–903; Galen A. McKinley et al., "Convergence of Atmospheric and North Atlantic Carbon Dioxide Trends on Multidecadal Timescales," *Nature Geoscience* 4 (June 2011): 606–10.

18. Intergovernmental Panel on Climate Change (IPCC), *Climate Change 2014: Impacts, Adaptation and Vulnerability. Contribution of Working Group II to the Fifth Assessment Report of the Intergovernmental Panel on Climate Change* (Cambridge, U.K. and New York: Cambridge University Press, 2014); NOAA Fisheries, *Ecosystem Advisory for the Northeast Shelf Large Marine Ecosystem*, no. 1 (2014).

19. IPCC, *Climate Change 2014: Impacts, Adaptation and Vulnerability*; Janet Nye et al., "Changing Spatial Distribution of Fish Stocks in Relation to Climate and Population Size on the Northeast United States Continental Shelf," *Marine Ecology Progress Series* 393 (2009): 111–29.

20. Rachel W. Obbard et al., "Global Warming Releases Microplastic Legacy Frozen in Arctic Sea Ice," *Earth's Future* 2, no. 6 (June 2014): 315–20.

21. Kara Lavender Law and Richard C. Thompson, "Microplastics in the Seas," *Science* 345, no. 6193 (July 11, 2014): 144–45.

22. Obbard et al., "Global Warming Releases Microplastic Legacy Frozen in Arctic Sea Ice."

23. NOAA Pacific Marine Environmental Laboratory (PMEL) Carbon Program, "What Is Ocean Acidification?" www.pmel.noaa.gov/co2/story/What+is+Ocean+Acidification%3F, viewed October 13, 2014; IPCC, *Climate Change 2014: Impacts, Adaptation and Vulnerability.*

24. NOAA PMEL Carbon Program, "What Is Ocean Acidification?"; IPCC, *Climate Change 2014: Impacts, Adaptation and Vulnerability.*

25. IPCC, *Climate Change 2014: Impacts, Adaptation and Vulnerability*; Victoria J. Fabry et al., "Ocean Acidification at High Latitudes: The Bellwether," *Oceanography* 22, no. 4 (2009): 160–71.

26. IPCC, *Climate Change 2014: Impacts, Adaptation and Vulnerability.*

27. Ibid.

28. Ibid.

29. Katherine E. Mills et al., "Fisheries Management in a Changing Climate: Lessons from the 2012 Ocean Heat Wave in the Northwest Atlantic," *Oceanography* 26, no. 2 (2013): 191–95; Michael Wines and Jess Bidgood, "Waters Warm, and Cod Catch Ebbs in Maine," *New York Times*, December 14, 2014; Pew Charitable Trusts, *Risky Decisions: How Denial and Delay Brought Disaster to New England's Historic Fishing Grounds*, 11.

30. "Atlantic Puffin Population Is in Danger, Scientists Warn," *Associated Press*, June 3, 2013.

31. Ibid.

32. IPCC, *Climate Change 2014: Impacts, Adaptation and Vulnerability*; Northern Economics, *The Seafood Industry in Alaska's Economy: 2011 Executive Summary Update* (Juneau, AK: Marine Conservation Alliance, 2011).

33. J. T. Mathis et al., "Ocean Acidification Risk Assessment for Alaska's Fishery Sector," *Progress in Oceanography* (2014), doi:10.1016/j.pocean.2014.07.001; IPCC, *Climate Change 2014: Impacts, Adaptation and Vulnerability.*

Chapter 7. Whose Arctic Is It?

1. Henry Huntington and Gunter Weller, "An Introduction to the Arctic Climate Impact Assessment," in *Arctic Climate Impact Assessment* (Cambridge, U.K.: Cambridge University Press, 2005), 3; 4 degrees from National Snow and Ice Data Center (NSIDC), "Climate Change in the Arctic," https://nsidc.org/cryosphere/arctic-meteorology/climate_change.html; 50-year period and Figure 7–1 from U.S. National Aeronautics and Space Administration, Goddard Institute for Space Studies, "GISS Surface Temperature Analysis (GISTEMP)," http://data.giss.nasa.gov/gistemp/maps/, viewed November 7, 2014.

2. Figure 7–2 from F. Fetterer et al., "Sea Ice Index," NSIDC, http://nsidc.org/data/seaice_index/, viewed November 2014; D. Perovich et al., "Sea Ice," in M. O. Jeffries, J. Richter Menge, and J. E. Overland, eds., *Arctic Report Card 2014* (Washington, DC: National Oceanic and Atmospheric Administration, December 2014), 35.

3. Kristina Pistone, Ian Eisenman, and V. Ramanathan, "Observational Determination of Albedo Decrease Caused by Vanishing Arctic Sea Ice," *Proceedings of the National Academy of Sciences* 111, no. 9 (2014): 3322–26.

4. See, for example, Natalia Shakhova et al., "Ebullition and Storm-Induced Methane Release from the East Siberian Arctic Shelf," *Nature Geoscience* 7 (2014): 64–70.

5. Arctic Monitoring and Assessment Programme (AMAP), *Arctic Ocean Acidification 2013: An Overview* (Oslo, Norway: 2014), ix.

6. Arctic Climate Impact Assessment, *Arctic Climate Impact Assessment* (Cambridge, U.K.: Cambridge University Press, 2005).

7. Jessica Gordon, "Inter-American Commission on Human Rights to Hold Hearing After Rejecting Inuit Climate Change Petition," *Sustainable Development Law & Policy*, Winter 2007, 55.

8. E. Rignot et al., "Acceleration of the Contribution of the Greenland and Antarctic Ice Sheets to Sea Level Rise," *Geophysical Research Letters* (2011), 38.

9. See, for example, Dim Coumou et al., "Quasi-Resonant Circulation Regimes and Hemispheric Synchronization of Extreme Weather in Boreal Summer," *Proceedings of the National Academy of Sciences* 111, no. 34 (August 26, 2014): 12331–36; Baek-Min Kim et al., "Weakening of the Stratospheric Polar Vortex by Arctic Sea-Ice Loss," *Nature Communications* 5 (September 2, 2014).

10. Save the Arctic website, www.savethearctic.org; European Commission, "European Parliament Resolution of 9 October 2008 on Arctic Governance" (Brussels: October 9, 2008).

11. Oklaik Eegeeisak, Chair, Inuit Circumpolar Conference, remarks made at Arctic Council meeting, Reykjavik, Iceland, November 2, 2014.

12. United Nations Convention on the Law of the Sea, Article 76 (5), Part VI, "Continental Shelf," www.un.org/depts/los/convention_agreements/texts/unclos/part6.htm.

13. Figure 7–3 from maribus gGmbH, *World Ocean Review* 1 (Hamburg: 2010).

14. Alaska Native Claims Settlement Act, www.law.cornell.edu/uscode/text/43/chapter-33; M. Nuttall, "Self-Rule in Greenland: Towards the World's First Independent Inuit State?" *Indigenous Affairs* 3–4 (2008): 64–70; Inuit Tapiriit Kanatami, "Vision of Self-Government," https://www.itk.ca/about-inuit/vision-self-government; Yukon First Nations Land Claims Settlement Act, http://laws-lois.justice.gc.ca/eng/acts/Y-2.3/; Canada Yukon Oil and Gas Accord, https://www.aadnc-aandc.gc.ca/eng/1369314748335/1369314778328; Yukon Northern Affairs Program Devolution Transfer Agreement, www.aadnc-aandc.gc.ca/eng/1297283624739/1297283711723; Northwest Territories Lands and Resources Devolution Agreement, http://devolution.gov.nt.ca/wp-content/uploads/2013/09/Final-Devolution-Agreement.pdf.

15. Robert Papp, Jr., "The U.S. Arctic Council Chairmanship," remarks made at the Passing the Arctic Council Torch conference, Centre for Strategic and International Studies, Washington, DC, September 30, 2014.

16. Clifford Krauss, "Shell Submits a Plan for New Exploration of Alaskan Arctic Oil," *New York Times*, August 28, 2014; Kevin McGwin, "Cairn 'Too Busy' for Greenland in 2014," *Arctic Journal*, January 21, 2014.

17. Gro Harlem Brundtland and World Commission on Environment and Development, *Our Common Future: Report of the World Commission on Environment and Development* (Oxford, U.K.: Oxford University Press, 1987), 41.

Chapter 8. Emerging Diseases from Animals

1. Sylvain Baize et al., "Emergence of Zaire Ebola Virus Disease in Guinea," *New England Journal of Medicine* 371 (October 9, 2014): 1418–25.

2. World Health Organization (WHO), "Situation Reports: Ebola Response Roadmap," www.who.int/csr/disease/ebola/situation-reports.en/; E. M. Leroy et al., "Multiple Ebola Virus Transmission Events and Rapid Decline of Central African Wildlife," *Science* 303, no. 5658 (January 16, 2004): 387–90; Heinz Feldmann and Thomas W. Geisbert, "Ebola Haemorrhagic Fever," *The Lancet* 377, no. 9768 (March 5, 2011): 849–62; Leroy et al., "Multiple Ebola Virus Transmission Events and Rapid Decline of Central African Wildlife."

3. WHO, "Marburg Virus Disease – Uganda," October 10, 2014, www.who.int/csr/don/10-october-2014-marburg/en/; Feldmann and Geisbert, "Ebola Haemorrhagic Fever."

4. International Livestock Research Institute, *Mapping of Poverty and Likely Zoonoses Hotspots*, Zoonoses Project 4, report to U.K. Department for International Development (Nairobi: 2012).

5. Kate E. Jones et al., "Global Trends in Emerging Infectious Diseases," *Nature* 451 (February 21, 2008): 990–93; J. Newcomb, T. Harrington, and S. Aldrich, *The Economic Impact of Selected Infectious Disease Outbreaks* (Cambridge, MA: Bio Economic Research Associates, 2011); Mark S. Smolinski, Margaret A. Hamburg, and Joshua Lederberg, *Committee on Emerging Microbial Threats to Health in the 21st Century, Microbial Threats to Health: Emergence, Detection, and Response* (Washington, DC: The National Academies Press, 2003).

6. Box 8–1 from U.S. Centers for Disease Control and Prevention, "Lesson 1: Introduction to Epidemiology," in Principles of Epidemiology in Public Health Practice, Third Edition. An Introduction to Applied Epidemiology and Biostatistics, Self-Study Course SS1978, www.cdc.gov/ophss/csels/dsepd/SS1978/Lesson1/Section10.html.

7. D. T. Haydon et al., "Identifying Reservoirs of Infection: A Conceptual and Practical Challenge," *Emerging Infectious Diseases* 8, no. 12 (December 2002): 1468–73; J. R. C. Pulliam et al. and the Henipavirus Ecology Research Group (HERG), "Agricultural Intensification, Priming for Persistence and the Emergence of Nipah Virus: A Lethal Bat-borne Zoonosis, *Journal of The Royal Society Interface* 9, no. 66 (January 2012): 89–101.

8. John Ford, *The Role of Trypanosomiasis in African Ecology* (Oxford, U.K.: Clarendon Press, 1971).

9. B. Hjelle and G. E. Glass, "Outbreak of Hantavirus Infection in the Four Corners Region of the United States in the Wake of the 1997–1998 El Nino-southern Oscillation," *The Journal of Infectious Diseases* 181, no. 5 (May 2000): 1569–73.

10. Todd R. Callaway et al., "Diet, *Escherichia coli* O:157:H7, and Cattle: A Review After 10 Years," *Current Issues in Molecular Biology* 11, no. 2 (2009): 67–79.

11. David A. Relman, "Microbial Genomics and Infectious Diseases," *New England Journal of Medicine* 365 (July 28, 2011): 347–57; A. R. Manges et al., "Comparative Metagenomic Study of Alterations to the Intestinal Microbiota and Risk of Nosocomial *Clostridum difficile*-associated Disease, *The Journal of Infectious Diseases* 202, no. 12 (December 15, 2010): 1877–84; M. Crhanova et al., "Immune Response of Chicken Gut to Natural Colonization by Gut Microflora and to *Salmonella enterica* Serovar Enteritidis Infection, *Infection and Immunity* 79, no. 7 (July 2011): 2755–63; Relman, "Microbial Genomics and Infectious Diseases."

12. Christopher Delgado et al., *Livestock to 2020: The Next Food Revolution,* Food, Agriculture, and the Environment Discussion Paper (Washington, DC: International Food Policy Research Institute, 1999).

13. Richard Coker et al., "Towards a Conceptual Framework to Support One-Health Research for Policy on Emerging Zoonoses, *The Lancet Infectious Diseases* 11, no. 4 (April 2011): 326–31; Delgado et al., *Livestock to 2020: The Next Food Revolution*; D. H. Molyneux, "Control of Human Parasitic Disease: Context and Overview, *Advances in Parasitology* 61 (2006): 1–43; WHO, Interagency Meeting on Planning the Prevention and Control of Neglected Zoonotic Diseases (NZDs), Geneva, Switzerland, July 5–6, 2011.

14. The Writing Committee of the WHO Consultation on Human Influenza A/H5, "Avian Influenza A (H5N1) Infection in Humans," *New England Journal of Medicine* 353 (September 29, 2005): 1374–85; William B. Karesh et al., "Wildlife Trade and Global Disease Emergence," *Emerging Infectious Diseases* 11, no. 7 (July 2005): 1000–02.

15. Trichinosis from D. G. Newell et al., "Food-borne Diseases – The Challenges of 20 Years Ago Still Persist While New Ones Continue to Emerge," *International Journal of Food Microbiology* 139, Supplement 1 (May 30, 2010): S3–15; 50 million from International Livestock Research Institute, *Mapping of Poverty and Likely Zoonoses Hotspots*; WHO, "WHO Consultation to Develop a Strategy to Estimate the Global Burden of Foodborne Diseases (Geneva: 2006); L. D. Sims et al., "Origin and Evolution of Highly Pathogenic H5N1 Avian Influenza in Asia," *Veterinary Record* 157, no. 6 (August 6, 2005): 159–64.

16. Karesh et al., "Wildlife Trade and Global Disease Emergence"; chimpanzees from Beatrice H. Hahn et al., "AIDS as a Zoonosis: Scientific and Public Health Implications, *Science* 287, no. 5453 (January 28, 2000): 607–14; Y. Guan et al., "Isolation and Characterization of Viruses Related to the SARS Coronavirus from Animals in Southern China," *Science* 302, no. 5643 (October 10, 2003): 276–78; P. Rouquet et al., "Wild Animal Mortality Monitoring and Human Ebola Outbreaks, Gabon and Republic of Congo, 2001–2003," *Emerging Infectious Diseases* 11, no. 2 (February 2005): 283–90.

17. J. A. Patz et al. and the Working Group on Land Use Change and Disease Emergence, "Unhealthy Landscapes: Policy Recommendations on Land Use Change and Infectious Disease Emergence," *Environmental Health Perspectives* 112, no. 10 (July 2004): 1092–98; A. Marm Kilpatrick and Sarah E. Randolph, "Drivers, Dynamics, and Control of Emerging Vector-borne Zoonotic Diseases," *The Lancet* 380, no. 9857 (December 1, 2012): 1946–55; P. S. Craig and the Echinococcosis Working Group in China, "Epidemiology of Human Alveolar Echinococcosis in China," *Parasitology International* 55, Supplement (2006): S221–25.

18. Jones et al., "Global Trends in Emerging Infectious Diseases"; Patz et al. and the Working Group on Land Use Change and Disease Emergence, "Unhealthy Landscapes: Policy Recommendations on Land Use Change and Infectious Disease Emergence"; J. F. Walsh, D. H. Molyneux, and M. H. Birley, "Deforestation: Effects on Vector-borne Disease," *Parasitology* 106, Supplement S1 (January 1993): S55–75; yellow fever and leishmaniasis from Bruce A. Wilcox and Brett Ellis, "Forests and Emerging Infectious Diseases of Humans," *Unasylva* 57, no. 224

(2006): 11–18.

19. Wilcox and Ellis, "Forests and Emerging Infectious Diseases of Humans"; Karesh et al., "Wildlife Trade and Global Disease Emergence"; J. R. Poulsen et al., "Bushmeat Supply and Consumption in a Tropical Logging Concession in Northern Congo," *Conservation Biology* 23, no. 6 (December 2009): 1597–608; N. Pramodh, "Limiting the Spread of Communicable Diseases Caused by Human Population Movement," *Journal of Rural and Remote Environmental Health* 2, no. 1 (2003): 23–32.

20. Mirko S. Winkler et al., "Assessing Health Impacts in Complex Eco-epidemiological Settings in the Humid Tropics: Advancing Tools and Methods," *Environmental Impact Assessment Review* 30, no. 1 (January 2010): 52–61.

21. A. M. Bal and I. M Gould, "Antibiotic Stewardship: Overcoming Implementation Barriers," *Current Opinion in Infectious Diseases* 24, no. 4 (August 2011): 357–62; Bonnie M. Marshall and Stuart B. Levy, "Food Animals and Antimicrobials: Impacts on Human Health," *Clinical Microbiology Reviews* 24, no. 4 (October 2011): 718–33.

22. Heather K. Allen et al., "Call of the Wild: Antibiotic Resistance Genes in Natural Environments," *Nature Reviews Microbiology* 8 (April 2010): 251–59; V M. D'Costa, E. Griffiths, and G. D. Wright, "Expanding the Soil Antibiotic Resistome: Exploring Environmental Diversity," *Current Opinion in Microbiology* 10, no. 5 (October 2007): 481–89.

23. R. H. Gustafson and R. E. Bowen, "Antibiotic Use in Animal Agriculture," *Journal of Applied Microbiology* 83 (1997): 531–41; Mary D. Barton, "Antibiotic Use in Animal Feed and Its Impact on Human Health," *Nutrition Research Reviews* 13 (2000): 279–99.

24. Mary J. Gilchrist et al., "The Potential Role of Concentrated Animal Feeding Operations in Infectious Disease Epidemics and Antibiotic Resistance," *Environmental Health Perspectives* 115, no. 2 (February 2007): 313–16; Marshall and Levy, "Food Animals and Antimicrobials: Impacts on Human Health"; Andreas Voss et al., "Methicillin-resistant *Staphylococcus aureus* in Pig Farming," *Emerging Infectious Diseases* 11, no. 12 (December 2005): 1965–66; Allen et al., "Call of the Wild: Antibiotic Resistance Genes in Natural Environments"; H. Heuer, H. Schmitt, and K. Smalla, "Antibiotic Resistance Gene Spread Due to Manure Application on Agricultural Fields," *Current Opinion in Microbiology* 14, no. 3 (June 2011): 236–43; M. F. Davis et al., "An Ecological Perspective on U.S. Industrial Poultry Production: The Role of Anthropogenic Ecosystems on the Emergence of Drug-resistant Bacteria from Agricultural Environments," *Current Opinion in Microbiology* 14, no. 3 (June 2011): 244–50; Marshall and Levy, "Food Animals and Antimicrobials: Impacts on Human Health"; Maria Sjölund et al., "Dissemination of Multidrug-resistant Bacteria into the Arctic," *Emerging Infectious Diseases* 14, no. 1 (January 2008): 70–72.

25. Stephen S. Morse et al., "Prediction and Prevention of the Next Pandemic Zoonosis," *The Lancet* 380, no. 9857 (December 1, 2012): 1956–65.

26. Coker et al., "Towards a Conceptual Framework to Support One-Health Research for Policy on Emerging Zoonoses"; William B. Karesh and Robert A. Cook, "The Human-Animal Link, One World – One Health," *Foreign Affairs* 84 (July/August 2005): 38–50; David Molyneux et al., "Zoonoses and Marginalised Infectious Diseases of Poverty: Where Do We Stand?," *Parasites & Vectors* 4 (2011): 106.

27. James O. Lloyd-Smith et al., "Epidemic Dynamics at the Human-Animal Interface," *Science* 326, no. 5958 (December 4, 2009): 1362–6; Morse et al., "Prediction and Prevention of the Next Pandemic Zoonosis."

28. Assaf Anyamba et al., "Prediction of a Rift Valley Fever Outbreak," *Proceedings of the National Academy of Sciences* 106, no. 3 (January 20, 2009): 955–5; Morse et al., "Prediction and Prevention of the Next Pandemic Zoonosis."

29. Paul R. Torgerson and Calum N. L. Macpherson, "The Socioeconomic Burden of Parasitic Zoonoses: Global Trends," *Veterinary Parasitology* 182, no. 1 (November 24, 2011): 79–95.

30. Jones et al., "Global Trends in Emerging Infectious Diseases"; L. H. Taylor, S. M. Latham, and E. Mark, "Risk Factors for Human Disease Emergence," *Philosophical Transactions of the Royal Society B: Biological Sciences* 356, no. 1411 (July 29, 2001): 983–89; Simon J. Anthony et al., "A Strategy to Estimate Unknown Viral Diversity in Mammals," *mBio* 4, no. 5 (September 3, 2013), e00598–13; World Bank, *People, Pathogens and Our Planet. Volume 2, The Economics of One Health* (Washington, DC: June 2012).

31. Allen et al., "Call of the Wild: Antibiotic Resistance Genes in Natural Environments"; Gilchrist et al., "The

Potential Role of Concentrated Animal Feeding Operations in Infectious Disease Epidemics and Antibiotic Resistance"; Heuer, Schmitt, and Smalla, "Antibiotic Resistance Gene Spread Due to Manure Application on Agricultural Fields"; Marshall and Levy, "Food Animals and Antimicrobials: Impacts on Human Health"; A. Parisien et al. "Novel Alternatives to Antibiotics: Bacteriophages, Bacterial Cell Wall Hydrolases, and Antimicrobial Peptides," *Journal of Applied Microbiology* 104, no. 1 (January 2008): 1–13.

Chapter 9. Migration as a Climate Adaptation Strategy

1. Cecilia Tacoli, *Not Only Climate Change: Mobility, Vulnerability and Socio-economic Transformations in Environmentally Fragile Areas in Bolivia, Senegal and Tanzania*, Human Settlements Working Paper 28 (London: International Institute for Environment and Development, 2011); Foresight, *Migration and Global Environmental Change: Future Challenges and Opportunities*, final project report (London: Government Office for Science, 2011).

2. WJ. McG. Tegart, G. W. Sheldon, and D. C. Griffiths, eds., *Climate Change: The IPCC Impacts Assessment. Report of Working Group II to the Intergovernmental Panel on Climate Change* (Canberra: Australian Government Publishing Service, 1990); Jon Barnett and Michael Webber, *Accommodating Migration to Promote Adaptation to Climate Change* (Washington, DC: World Bank: 2010).

3. Hundreds of millions from Nicholas Stern, *The Economics of Climate Change. The Stern Review* (Cambridge, U.K.: Cambridge University Press, 2007); German Advisory Council on Global Change (WBGU), *Climate Change as a Security Risk* (London: Earthscan, 2008); Global Humanitarian Forum, *The Anatomy of A Silent Crisis* (Geneva: 2009); adaptation strategy from David Rain, *Eaters of the Dry Season: Circular Labor Migration in the West African Sahel* (New York: Westview Press, 1999), and from Kees van der Geest, *Migration and Natural Resources Scarcity in Ghana*, case study report for the EACH-FOR project (Brussels: EACH-FOR, 2009).

4. Norman Myers, "Environmental Refugees: A Growing Phenomenon of the 21st Century," *Philosophical Transactions of the Royal Society B* 357, no. 1420 (April 29, 2002): 609–13; Molly Conisbee and Andrew Simms, *Environmental Refugees. The Case for Recognition* (London: New Economics Foundation, 2003); Nicholas Stern, *The Global Deal. Climate Change and the Creation of a New Era of Progress and Prosperity* (New York: Public Affairs, 2009); migration as a resource from Rain, *Eaters of the Dry Season*, and from Richard Black et al., "Climate Change: Migration as Adaptation," *Nature* 478 (October 27, 2011): 447–49.

5. Carol Farbotko, "Wishful Sinking: Disappearing Islands, Climate Refugees and Cosmopolitan Experimentation," *Asia Pacific Viewpoint* 51, no. 1 (April 2010): 47–60; Jon Barnett and John Campbell, *Climate Change and Small Island States. Power, Knowledge and the South Pacific* (London: Earthscan, 2010).

6. Robert J. Nicholls et al., "Sea-level Rise and Its Possible Impacts Given a 'Beyond 4°C World' in the Twenty-first Century," *Philosophical Transactions of the Royal Society A* 369, no. 1934 (January 13, 2011): 161–81.

7. Postcards from the Future, "Buckingham Palace Shanty," October 4, 2010, www.london-futures.com/2010/10/04/buckingham-palace-shanty/.

8. WBGU, *Climate Change as a Security Risk*; European Commission and the Secretary-General/High Representative, *Climate Change and International Security* (Brussels: Council of the European Union, 2008).

9. Richard Black, *Environmental Refugees: Myth or Reality?* New Issues in Refugee Research, Working Paper No. 34 (Geneva: United Nations High Commissioner for Refugees, 2001); Camillo Boano et al., *Environmentally Displaced People: Understanding the Linkages Between Environmental Change, Livelihoods and Forced Migration* (Oxford, U.K.: Refugee Studies Centre, 2007).

10. Stefan Rahmstorf, "A New View on Sea Level Rise," *Nature Reports Climate Change* 4 (2010): 44–45.

11. Gunvor Jónsson, *The Environmental Factor in Migration Dynamics – A Review of African Case* Studies, Working Papers No. 21 (Oxford, U.K.: International Migration Institute, 2010); Black, *Environmental Refugees: Myth or Reality?*; van der Geest, *Migration and Natural Resources Scarcity in Ghana.*

12. Box 9–1 from the following sources: Internal Displacement Monitoring Centre, *Global Estimates 2014. People Displaced by Disasters* (Geneva: September 2014), 8, 15, 36–38; Figure 9–1 from idem; short-distance, temporary from Frank Laczko and Christine Aghazarm, eds., *Migration, Environment and Climate Change: Assessing the*

Evidence (Geneva: International Organization for Migration, 2009), 23; Hurricane Katrina from Fabrice Renaud et al., *Adapt or Flee. How to Face Environmental Migration?* InterSecTions No. 5 (Bonn: United Nations University Institute for Environment and Human Security, May 2007), 22.

13. François Gemenne, "What's in a Name: Social Vulnerabilities and the Refugee Controversy in the Wake of Hurricane Katrina," in Jill Jäger and Tamer Afifi, *Environment, Forced Migration and Social Vulnerability* (Berlin: Springer, 2010); Barnett and Webber, *Accommodating Migration to Promote Adaptation to Climate Change.*

14. Jill Jäger et al., *EACH-FOR Synthesis Report* (Budapest: EACH-FOR, 2009).

15. R. McLeman and B. Smit, "Migration as an Adaptation to Climate Change," *Climatic Change* 76, no. 1-2 (May 2006): 31–53; Matthew Walsham, *Assessing the Evidence: Environment, Climate Change and Migration in Bangladesh* (Geneva: International Organization for Migration, 2010); Jeni D. Klugman, *Human Development Report 2009. Overcoming Barriers: Human Mobility and Development* (New York: United Nations Development Programme, 2009).

16. Development Research Centre on Migration, *Globalisation and Poverty, Making Migration Work for Development* (Brighton, U.K.: University of Sussex, 2009); McLeman and Smit, "Migration as an Adaptation to Climate Change."

17. François Gemenne, "Climate-induced Population Displacements in a 4°C+ World," *Philosophical Transactions of the Royal Society A* 369, no. 1934 (2011): 182–95.

18. Graeme Hugo, "Environmental Concerns and International Migration," *International Migration Review* 30, no. 1 (Spring 1996): 105–31.

19. Gemenne, "Climate-induced Population Displacements in a 4°C+ World."

Chapter 10. Childhood's End

1. Albert O. Hirschman, *Exit, Voice, and Loyalty: Responses to Decline in Firms, Organizations, and States* (Cambridge, MA: Harvard University Press, 1970).

2. Edward O. Wilson, *The Meaning of Human Existence* (New York: Liveright, 2014), 120.

3. Intergovernmental Panel on Climate Change, *Climate Change 2014 Synthesis Report: Summary for Policymakers* (Cambridge, U.K. and New York: Cambridge University Press, 2014), 1–4.

4. Lenore Taylor and Tania Branigan, "US and China Strike Deal on Carbon Cuts in Push for Global Climate Change Pact," *The Guardian* (U.K.), November 12, 2014; Harald Winkler, "How Long Can You Go? Climate Talks in Lima," University of Cape Town Energy Research Centre blog, December 15, 2014, www.ercblogs.co.za/2014.

5. See Herman E. Daly, *Beyond Growth: The Economics of Sustainable Development* (Boston: Beacon Press, 1996), 6.

6. "Five Questions for Gus Speth on His Environmental Evolution," *Yale Environment 360*, December 2, 2014, http://e360.yale.edu.

7. Joan Gussow and Katherine Clancy, "Dietary Guidelines for Sustainability," *Journal of Nutrition Education* 18, no. 1 (February 1986): 1; Dan Charles, "Congress to Nutritionists: Don't Talk About the Environment," National Public Radio blog, December 15, 2014, www.npr.org/blogs/thesalt/.

8. A more detailed discussion of panarchy theory can be found in Thomas Homer-Dixon, *The Upside of Down: Catastrophe, Creativity, and the Renewal of Civilization* (Washington, DC: Island Press, 2006), 225–234.

9. Herman E. Daly, *Steady-State Economics*, 2nd edition (Washington, DC: Island Press, 1991), 45.

10. Worldwatch Institute, *State of the World 2014: Governing for Sustainability* (Washington, DC: Island Press, 2014), 249.

11. Tom Sanzillo et al., *Material Risks: How Public Accountability Is Slowing Tar Sands Development* (Washington, DC: Oil Change International and Institute for Energy Economics & Financial Analysis, October 2014).

12. See Tina Nabatchi et al., *Democracy in Motion: Evaluating the Practice and Impact of Deliberative Civic Engagement* (Oxford, U.K.: Oxford University Press, 2012); James Fishkin, "Consulting the Public Thoughtfully: Prospects for Deliberative Democracy," in David Kahane et al., eds., *Deliberative Democracy in Practice* (Vancouver, BC: UBC Press, 2010), 194.

13. Matt Leighninger, "The Next Form of Democracy?" the 2012 Civic Engagement and Democracy Lecture, delivered at the Institute for Policy and Civic Engagement, University of Illinois–Chicago, Chicago, IL, April 4, 2012.

14. Ibid.

15. See Richard Wilkonson and Kate Pickett, *The Spirit Level: Why Greater Equality Makes Societies Stronger* (New York: Bloomsbury Press, 2009).

16. Box 10–1 from the following sources: production for surplus from John Gowdy, "Governance, Sustainability, and Evolution," in Worldwatch Institute, *State of the World 2014: Governing for Sustainability*, 31–40; for a striking exception to the rule of elites over masses, the Swiss canton of Graubünden, see Benjamin Barber, *The Death of Communal Liberty: A History of Freedom in a Swiss Mountain Canton* (Princeton, NJ: Princeton University Press, 1974); Ronald Wright, *A Short History of Progress* (New York: Carroll & Graf, 2005), 51.

Index

Page numbers in *italics* indicate illustrations, tables, and figures.

Island Press | Board of Directors